A USER'S GUIDE TO ENPORT-4

Professor Ronald C. Rosenberg
Department of Mechanical Engineering
Michigan State University
East Lansing, Michigan

with the assistance of Mark McPherson

July 1974

A WILEY-INTERSCIENCE PUBLICATION

JOHN WILEY & SONS, New York • London • Sydney • Toronto

Library of Congress Cataloging in Publication Data:

Rosenberg, Ronald C.
A user's guide to ENPORT-4.

"A Wiley Interscience publication."
1. Bond graphs. 2. ENPORT-4 (Computer program)
3. Digital computer simulation. I. McPherson, Mark, joint author. II. Title.

TA338.B6R67 621'.01'84 74-19411
ISBN 0-471-73614-7

Printed in the United States of America

10 9 8 7 6 5 4 3 2 1

4.95

PREFACE

A large number of physical and engineering systems may be represented directly in terms of component energy characteristics and their power interactions. When the system elements are modeled as energetic multiports and their interconnections by power bonds, then the bond graph language is a natural one for describing the entire system. Bond graphs may be written for dynamic systems involoving various energy types, such as electrical, mechanical, fluid, and thermal; all energy types may be coexistent. Useful modeling elements include multiport storages, dissipators, and junction elements and transducers, as well as sources.

Bond graph models of linear multiport systems may be transformed to state-space form by a powerful algorithm based upon operational causality. From the state-space equations dynamic responses may be obtained by the matrix exponential technique, thereby allowing the direct digital simulation of linear multiport models.

The ENPORT-4 program is a realization of the bond graph linear reduction algorithm. It is a principal purpose of this guide to describe the procedure upon which ENPORT-4 is based and to present some examples of its use. Important features of ENPORT-4 are its choice of physically-significant state variables, its use of operational causality to obtain an orderly formulation of system equations, and its ability to handle systems containing static storage subfields.

The development of more efficient, more powerful ENPORT programs is a continuing effort. The current version, ENPORT-4, will remain a standard program as described in this guide. Subsequent versions are planned to accomodate an engineering device library facility and nonlinear elements. These will appear as ENPORT-5 and ENPORT-6, respectively.

After a number of years of effort in ENPORT-related activities I find my interest still keen and the potential contributions of the approach still largely unexplored and unrealized. To those whose curiousity and enthusiasm are stimulated by their experiences with the program or its underlying concepts I extend an invitation to engage in further discussion.

ACKNOWLEDGEMENTS

The author wishes to acknowledge the major influence that Professor Henry M. Paynter, inventor of the bond graph language, has had on this work. It was he who first proposed the creation of an n-port, or ENPORT, program (in 1963), and his early suggestions were helpful in guiding the work through several stages of evolution.

Mark McPherson has played a key role in implementing version 4 of ENPORT on the CDC 6500 system at Michigan State University in both batch and interactive modes. His contributions have been essential.

And to my long-time friend and colleague, Professor Dean Karnopp, must go credit for trying to keep the ENPORT program "reality-oriented" in the development of several versions. The user of ENPORT-4 may be the judge of his success.

The Computer Laboratory at Michigan State University has been very cooperative in providing program file maintenance, as well as the inevitable debugging assistance required in getting any large, complex program to run reliably. The College of Engineering has assisted in many ways, most notably in providing programming assistance and distribution support.

CONTENTS

CONTENTS (continued)

CHAPTER 1. INTRODUCTION

In our technological world each new year brings an increase in the size and complexity of the physical and engineering systems conceived, designed, analyzed, and implemented. On occasion there are behaviors of a complex system that have escaped notice in the preliminary design stages, only to emerge in operation, producing such undersirable effects as electric power blackout or "pogo" vibrations in a rocket. To maintain control over these complex creations that is as effective as possible, it is necessary to develop tools for modeling, analysis, and simulation equal to the challenge. This guide describes a powerful tool--the ENPORT-4 program--designed to achieve for the engineer an effective partnership with the computer in the study of large-scale, linear, multiport systems.

The three principal steps to be taken by the engineer-computer combination are:

(1) To develop a mathematical model in some suitable form;

(2) To select a formulation, typically in terms of state-space differential equations; and

(3) To effect a calculation based on the formulation that yields the dynamic behavior of desired response variables.

In the procedure outlined in this guide, the first step is taken by the engineer, who develops a multiport model in bond graph terms. Descriptions of such models are given in section 1. The next two steps of the procedure are performed by the computer, which selects state variables of physical significance, formulates a set of state-space equations, and calculates the response indicated by the formulation.

1.1 Purpose of the Guide

The User's Guide to ENPORT-4 has three principal purposes. These are:

(1) to introduce the description of dynamic system models in terms suitable for the ENPORT-4 program to process them;

(2) to specify the mechanics of operation of ENPORT-4 in detail, so that useful results may be obtained; and

(3) to suggest, by means of examples, some strategies of modeling and investigation useful in studying dynamic system behavior.

It is not the purpose of this guide to show how to construct bond graph models of multiport systems; for the inexperienced bond grapher some helpful information about references is given in Section 1.5. Also, a concise definition of the bond graph language is included in this guide as Appendix A. The guide is viewed as a working companion to the book <u>System Dynamics</u> by D.C. Karnopp and R.C. Rosenberg, Wiley-Interscience, 1975.

1.2 Organization of the Guide

The remainder of Chapter 1 is devoted to the general consideration of formulating dynamic system models in terms suitable for ENPORT-4. The next section (1.3) presents four examples suggestive of the wide range of applicability of the approach. Section 1.4 defines the parameters used by ENPORT-4 and shows their relation to common physical and engineering properties. In section 1.5 a thorough discussion of available reference material is given, relating both to bond graph modeling and to bond graph processing and formulation techniques.

Chapter 2 specifies in detail the information required for setting up and running batch (i.e., card input-printer output) jobs under standard ENPORT-4. The various commands and control data used in the examples of Chapter 1 are described there, as well as illustrations of various output result options.

In Chapter 3 the interactive (i.e., on-line) version of ENPORT-4 is described. Since its use is similar to that of the batch version, and is in fact somewhat simpler because of question-and-answer cues, the user familiar with the batch version should have little difficulty using interactive ENPORT-4.

Finally, in Chapter 4 the basic design of the ENPORT-4 program is described. For a somewhat experienced user, an understanding of the program design will permit a fuller utilization of ENPORT-4's potential, which this Guide can hint at, but not explain completely.

A set of appendices contains detailed information about the bond graph language and ENPORT-4 organization and operation. Questions regarding the use of the program may be addressed to the author, as follows:

Professor Ronald C. Rosenberg
Department of Mechanical Engineering
Michigan State University
East Lansing, Michigan 48824

1.3 Examples of Use

In this section four examples of the use of ENPORT-4 are given. Each is intended to illustrate some of the features of the program and of bond graph modeling. Detailed descriptions of the job deck, command usage, and other program options are reserved to chapters 2 and 3. Those examples should be scanned for overall style and pattern and may be used as test cases for checking your understanding of program use.

A serious study of the dynamics of any of the example systems would require a number of passes or jobs to be run, as key parameters and conditions are varied. The single pass for each example should be viewed as suggestive, not definitive (as any experienced system dynamicist would agree).

1.3.1. A hydraulic system example

This example illustrates the basic pattern of making a bond graph model of a physical or engineering system, describing the model to ENPORT-4, and obtaining results for confirming or altering the design of the system.

Figure 1-1(a) shows a two-tank, two-valve hydraulic system with an inflow (source) and an outflow. In part (b) of Figure 1-1 a bond graph model for the hydraulic system is given. The capacitance elements (C2 and C6) represent the tanks, the resistance elements (R4 and R7) represent the valves, and the inflow is represented by the flow source element (SF 1). The 0-junctions represent common effort (pressure connections) and the 1-junction represents the common flow connection between the two tanks through the valve.

ENPORT-4 works with generalized power variables e(effort) and f(flow) and their time integrals p (momentum) and q (displacement), respectively. In this hydraulic example we identify the variables and choose their units as follows:

ENPORT-4 Variable	physical Variable	assumed Units
E	P, pressure	[lbf/ft^2]
F	Q, flow	[ft^3/sec]
P	T, pressure momentum	[lbf-sec/ft^2]
Q	V, volume	[ft^3]

Referring to Figure 1-2, the first two cards specify the run label for this job. The next three cards are used to specify the bond graph of Figure 1-1(b). Each element is listed, immediately followed by its bonds. This is called the line code for the graph.

In the hydraulic system we assume the following values for physical and engineering constants:

Area of each tank = 3.2 [ft^2]
Valve resistances = 1170. [lbf-sec/ft^5]
Working fluid (water) = 62.4 [lbf-ft^3]

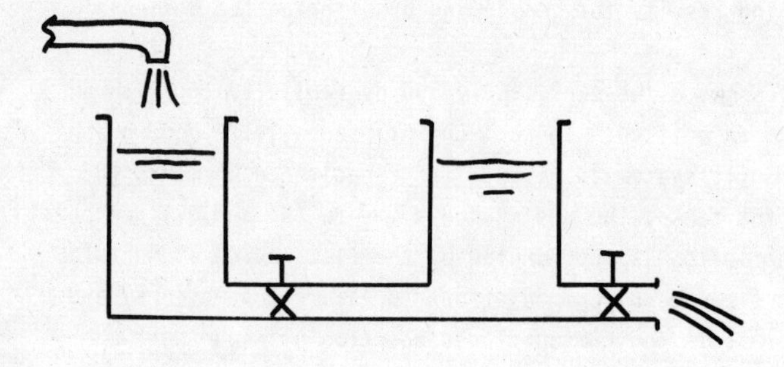

(a) hydraulic system

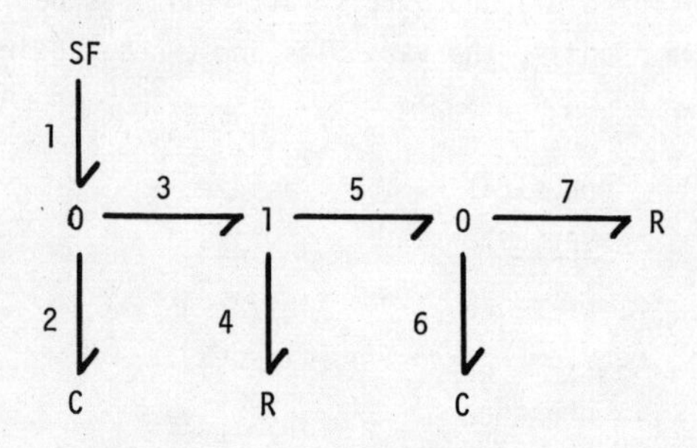

(b) bond graph model for (a)

Figure 1-1. A hydraulic system example.

```
M S U   6 5 0 0   ENPORT                     VERSION 4.1             UPDATE 50772
--------------------------------------------------------------------------------


CONTENTS OF TAPE 2   (INPUT)

CARD  1 = {HEADER
CARD  2 = {HYDRAULIC SYSTEM -- FIGURE 1-1.
CARD  3 = {GRAPH
CARD  4 = {SF 1 , C 2 , 0 1 2 3 , 1 3 4 5 , R 4 , 0 5 6 7 , C 6 , R 7.
CARD  5 = {GREND
CARD  6 = {PARAMETERS
CARD  7 = {C2=  19.5
CARD  8 = {R4= 1170.
CARD  9 = {C6= 19.5
CARD 10 = {R7= 1170.
CARD 11 = {PREND
CARD 12 = {SOURCES
CARD 13 = {F1=  CONSTANT 0.187
CARD 14 = {SREND
CARD 15 = {SOLUTION
CARD 16 = {TIN= 0.0   TFIN= 200.   DT= 0.5.
CARD 17 = {ICS= 3.2 3.2.
CARD 18 = {TPLOT 0.0 200. 8 Q2 Q6.
CARD 19 = {SLEND
CARD 20 = {JOBEND

END-OF-FILE
```

Figure 1-2. ENPORT-4 job deck for the hydraulic system example.

These data may be converted to the appropriate values for the C- and R-element parameters, using the information given in section 1.4. The parameters are specified by cards 6 to 11 in Figure 1-2.

The inflow source is specified by cards 12, 13, and 14 as constant, with a value of 0.187 [ft^3/sec].

Finally, the necessary data to control, initialize, and display results are specified on cards 15 to 19. The run will start at time 0.0 (seconds), end at time 200. (seconds), and results will be saved at intervals of 0.5 (seconds). The initial conditions (volumes of water in tanks 2 and 6 respectively) are 3.2 [ft^3] and 3.2 [ft^3], which correspond to heights of 1.0 [ft]. The results will be plotted (on the printer) from time 0.0 to 200.0 in steps of 4.0 seconds (i.e., 8 times 0.5); the variables are volumes in the two tanks. (Remember, the Q's are generalized displacement variables in ENPORT-4.) The generated plot is shown in Figure 1-3, and indicates that the two volumes (hence the two levels) rise to new steady values which allow for the constant inflow to pass through the system.

If changes to system parameters or the initial conditions or the inflow were desired, they could be specified next, or a new job could be run. Card 20 of Figure 1-2 terminates this particular run.

Summing up this example, we see that all ENPORT-4 required was the bond graph, parameter and source specification, and some simulation and results control data. There is no more compact way to study this system with the aid of a computer.

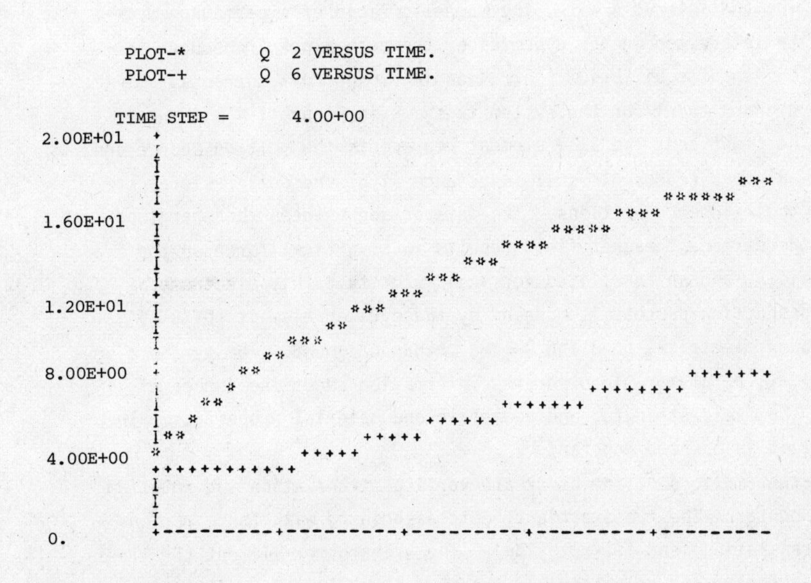

Figure 1-3. Plot of hydraulic system response showing volumes in the tanks.

1.3.2 A beam-block transducer system

This example is included to illustrate the use of ENPORT-4 in studying systems containing transducer effects and multiport field elements. Figure 1-4(a) depicts a block rigidly attached to a cantilevered beam. A coil is wound around the block and energized by a power source. The block-coil sits in, and interacts with, the magnetic field of a permanent magnet. We are interested in the dynamics of the beam-block transducer for small motions about equilibrium (corresponding to no coil current).

A bond graph model for the system is shown in Figure 1-4(b). Starting from the right end, the SE 7 element represents the voltage source and the 1 -I - R group represents coil inductance (I 5) and coil resistance (R6), and their interconnections. The interaction between the energized coil and the permanent magnet field results in a vertical force on the coil and block, and an associated vertical velocity. This electromechanical transduction process is modeled by the gyrator element (GY 8 9), where 8 is an electrical bond and 9 is a mechanical bond. The gyrator modulus (i.e., parameter) incorporates information about the number of windings, the field strength, and geometric and material properties. In this model it is assumed constant.

Important motions of the block are vertical translation and rotation about its center. The two inertia effects associated with these motions are I4 (translation) and I3 (rotation). The transformer element (TF 11 12) relates forces and velocities at the beam-block junction point to the force-velocity pair at the center of mass of the block. The beam itself can be viewed as a two-port compliance element, storing elastic energy due both to vertical deflection and to rotation at its free end, where the block is joined. The beam is represented by the two-port element C 1 2. Bond 1 has rotational power (torque times angular velocity) and bond 2 has translation power (force times velocity). For the interested reader a more complete discussion of this example is to be found in references [10] and [11] (see section 1.5). Here we shall concentrate on the pattern of ENPORT-4 usage.

The physical variables are identified as follows:

bonds 5,6,7,8 ----- electrical (voltage, current)

bonds 1,4,9,10,12 ----- mechanical translation (force, velocity)

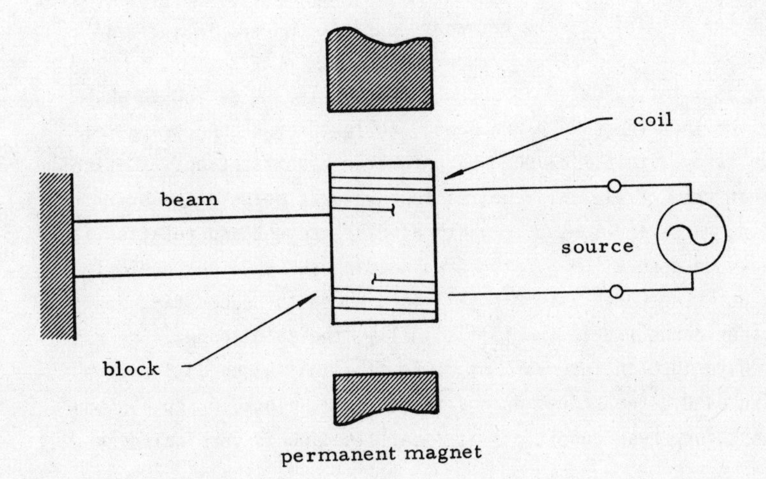

(a) the transducer schematic

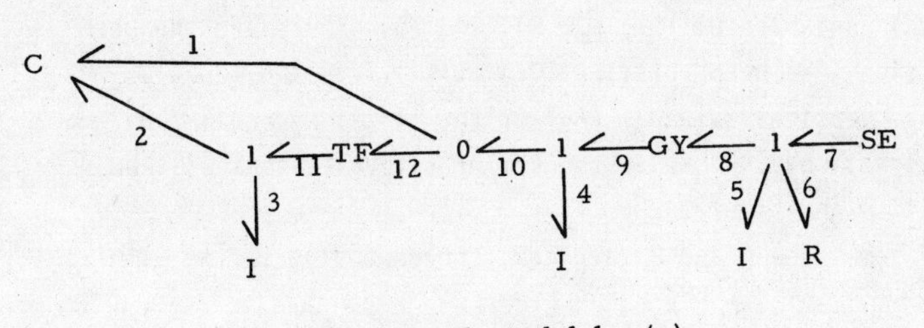

(b) bond graph model for (a)

Figure 1-4. A beam-block transducer system.

bonds 2,3,11 ----- mechanical rotation
(torque, angular velocity)

We shall not discuss the calculation of element parameters for this example, but again point out the necessary definitions are included in Section 1.4.

The job deck for ENPORT-4 is shown in Figure 1-5. Reading through the deck we see that there is a run label, followed by a bond graph description on cards 3 to 6, followed by parameter specifications. Notice on card 8 that the C12 element requires four values, being a two-by-two array relating force and torque to vertical displacement and rotation at the free end. On card 9 the gyrator transduction (GY 8 9) has a single, constant value. The voltage source (E7) is assumed to be constant at value 10.0 (see cards 11, 12 and 13). Finally, the solution will be run from time 0.0 to 10.0 in intervals of 0.05. The initial conditions are all specified as 0.0 (no stored energy in the beam, block, or coil), and a plot request for these conditions is made. The job is terminated by card 19.

Some intermediate processing results are shown in Figures 1-6 and 1-7. In Figure 1-6 the initial processing of the bond graph is displayed, including graph structure and bond power directions. Then the key variables for formulation are identified, and include the independent energy (or state) variables Q1, Q2, P3, P4, and P5. These are the beam vertical deflection (Q1), beam rotation (Q2) at the free end, angular momentum (P3) and vertical momentum (P4) of the block, and coil flux linkage (P5) respectively. This is the set of variables that will be used in formulation and simulation.

Figure 1-7 shows the A and B matrices, corresponding to the matrix state equation

$$\dot{X}(t) = A*X(t) + B*U(t),$$

where

$$X = \begin{bmatrix} Q1 \\ Q2 \\ P3 \\ P4 \\ P5 \end{bmatrix}, \text{ the state vector },$$

```
M S U   6 5 0 0   ENPORT              VERSION 4.1            UPDATE  50772
--------------------------------------------------------------------------

CONTENTS OF TAPE2  (INPUT)

CARD  1 = [HEADER
CARD  2 = [BEAM BLOCK TRANSDUCER SYSTEM.
CARD  3 = [GRAPH
CARD  4 = [C 1 2 , I 3 , I 4 , I 5 , R 6 , 1 7 8 5 6 , GY 8 9 ,
CARD  5 = [1 4 9 10 , 0 10 12 1 , TF 11 12 , 1 2 3 11 , SE 7 .
CARD  6 = [GREND
CARD  7 = [PARAMETERS
CARD  8 = [C12= 7.0 4.0 4.0 9.0  I3= 1.0  I4= 2.0  I5= 3.0
CARD  9 = [R6= 6.0  GY89= 5.0  TF1112= 1.5
CARD 10 = [PREND
CARD 11 = [SOURCES
CARD 12 = [E7= CONSTANT  10.0
CARD 13 = [SREND
CARD 14 = [SOLUTION
CARD 15 = [TIN= 0.0  TFIN= 10.0  DT=  0.05
CARD 16 = [ICS= 0.0 0.0 0.0 0.0 0.0 .
CARD 17 = [TPLOT 0.0 10.0 2 Q1 Q2 P4 .
CARD 18 = [SLEND
CARD 19 = [JOBEND

END-OF-FILE
```

Figure 1-5. ENPORT-4 job deck for the beam-block transducer system.

```
SYSTEM BOND GRAPH

NODE     BONDS
 C          1    2
 I          3
 I          4
 I          5
 R          6
 1          7    8    5    6
 GY         8    9
 1          4    9   10
 0         10   12    1
 TF        11   12
 1          2    3   11
 SE         7

GRAPH ANALYSIS.

 THE GRAPH HAS    12 NODES AND    12 BONDS.

 POWER FLOWS
BOND  FROM TO
  1    0  → C
  2    1  → C
  3    1  → I
  4    1  → I
  5    1  → I
  6    1  → R
  7    SE → 1
  8    1  → GY
  9    GY → 1
 10    1  → 0
 11    TF → 1
 12    0  → TF

THERE ARE    7 FIELD BONDS AND    5 JUNCTION BONDS.

THE BONDS INTERNAL TO THE JUNCTION STRUCTURE ARE --
 12 11 10  9  8

THE  5 INDEPENDENT ENERGY VARIABLES ARE --

Q 1  Q 2  P 3  P 4  P 5

THE  1 DISSIPATION VARIABLES ARE --

E 6

THE  1 SOURCE VARIABLES ARE --

E 7
```

Figure 1-6. ENPORT-4 analysis and set-up of the beam-block transducer system.

THE A MATRIX

0.	0.	-1.500E+00	2.000E+00	0.
0.	0.	1.000E+00	0.	0.
6.500E+00	-3.000E+00	0.	0.	0.
-7.000E+00	-4.000E+00	0.	0.	1.500E+01
0.	0.	0.	-1.000E+01	-1.800E+01

THE B MATRIX

0.
0.
0.
0.
1.000E+00

THE EIGENVALUES OF THE A MATRIX...

NUMBER	REAL PART	IMAGINARY PART	TIME CONSTANT
1	-8.0581E+00	8.2400E+00	-1.2410E-01
2	-8.0581E+00	-8.2400E+00	-1.2410E-01
3	-4.3948E-01	3.5333E+00	-2.2754E+00
4	-4.3948E-01	-3.5333E+00	-2.2754E+00
5	-1.0048E+00	0.	-9.9521E-01

THE STEADY STATE X-VECTOR IS...
.5319E+00 .1152E+01 .7562E-12 .7400E-12 .5556E+00

Figure 1-7. ENPORT-4 analysis and formulation of the beam-block transducer system.

and

$U = [E7]$, the source vector.

Insight into the system dynamics is available by inspecting the eigenvalues of A, printed next. There are two pairs of complex-conjugate roots and one real root, all with negative real parts. Therefore the system is stable, but it may oscillate.

Since the source is constant a steady-state is available as a set of constant values for X. With a constant voltage imposed, the system finally settles down to a position with the beam deflected and rotated, no motion of the block in either rotation or translation, and a constant coil flux linkage. These results are printed in Figure 1-7.

Finally, the plot of beam rotation (Q1), beam deflection (Q2), and block vertical momentum (P4) is shown in Figure 1-8. The dynamics are (or should be) no great suprise. Under the influence of a constant voltage the block rotates and translates upward in an oscillatory fashion, coming to rest in a deflected steady-state.

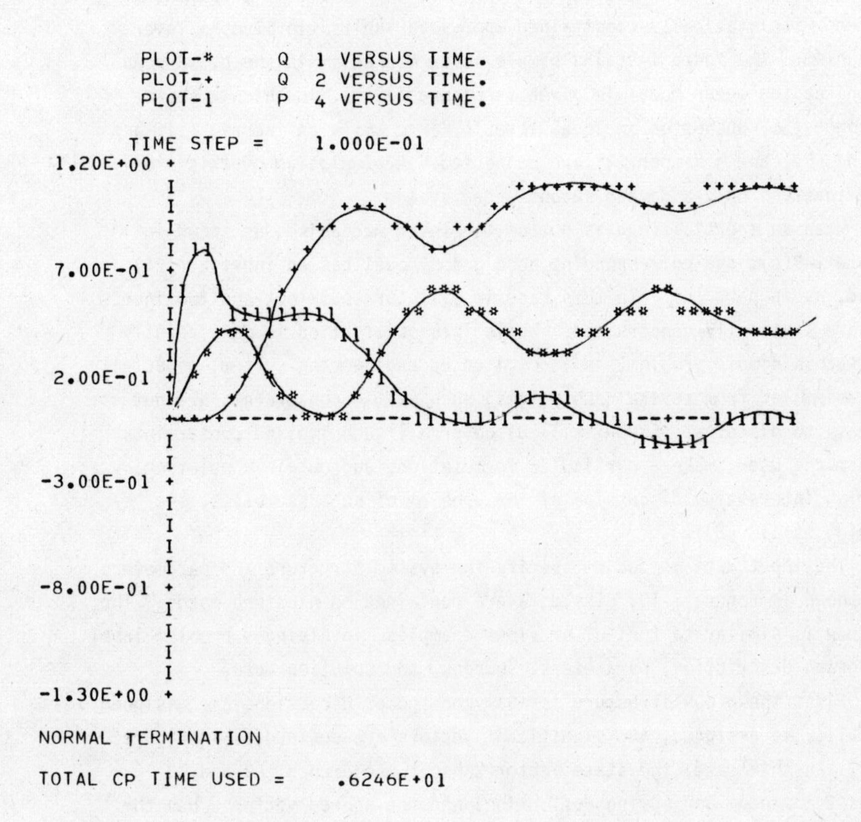

Figure 1-8. Plot of dynamic response of the beam-block transducer system.

1.3.3 A lever mechanism

A third example, intended to illustrate the automatic treatment of systems with statically-constrained energy variables, involves a lever mechanism. In Figure 1-9(a) a simple lever mechanism is shown. A bond graph for the lever model is given in Figure 1-9(b), in which a source of force (SE) pushes on an ideal lever (TF), to which an inertia (I), a spring (C), and a damper (R) are connected. A simulation of this problem involves only a damped second-order system.

When an inertial load is put on the lever mechanism, as shown in Figure 1-9(c), the corresponding bond graph model has an inertia (I 4) added, as in part (d). In this case it will turn out that the two inertias are statically-constrained. While this observation is made readily for such a simple problem, in a complicated engineering system, especially one assembled from several subsystems, such static constraints are not so easy to discover. ENPORT will discover all such implied constraints, inform the user, make a particular formulation, and obtain a solution. (For an interesting discussion of the problem of sub-assemblies, see Koenig, et al. [8]).

The input data needed to specify the system structure and parameters are shown in Figure 1-10. The data are contained on nineteen cards. The pattern is similar to that of previous examples, involving a problem label, the graph description, parameters, sources, and solution data.

First the graph structure is analyzed, power directions are assigned, causality is assigned, and significant vectors are defined as in Figure 1-11. In this case, the state vector X has two elements, P2 and Q3 (mass 2 momentum and spring deflection) and the source vector U has the single element E1 (force 1). Notice that P4 (momentum of I4 or mass added) is taken as dependent. Formulation proceeds with the resulting matrix equation implied by A, B and E:

$$\dot{X} = A*X + B*U + E*\dot{U}, \qquad [1\text{-}2]$$

shown in Figure 1-12, where the E matrix must be considered due to the static constraint among energy variables in the model. Also, in the same figure the eigenvalues of A are given, along with the steady state. The A matrix is 2x2, since X is second order. The steady state values

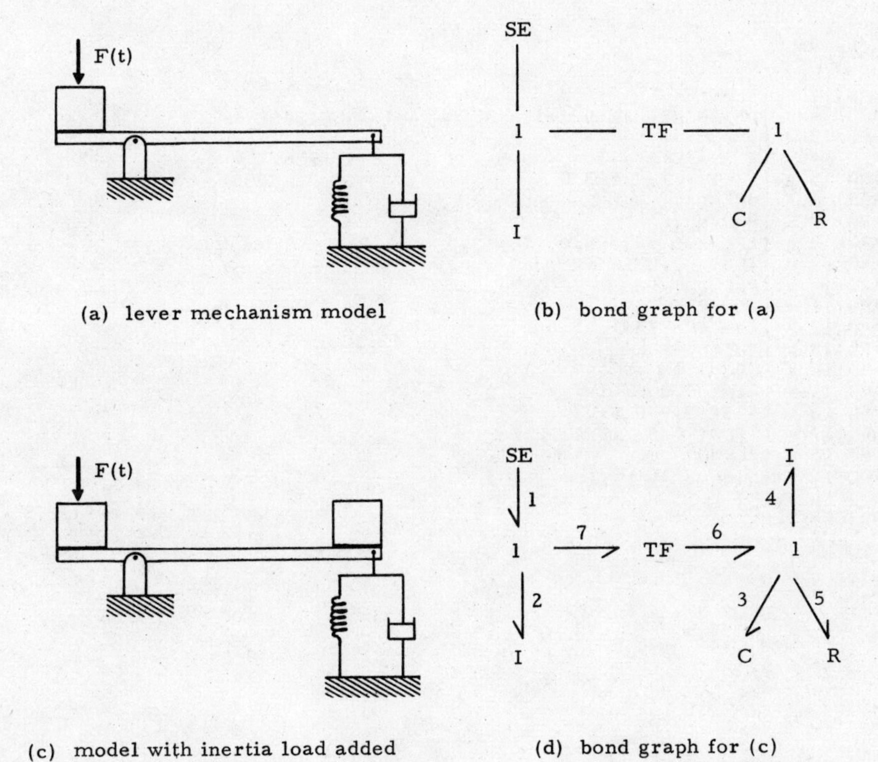

Figure 1-9. A lever mechanism with inertia load.

```
M S U   6 5 0 0   ENPORT          VERSION 4.1          UPDATE  50772
-----------------------------------------------------------------------


CONTENTS OF TAPE2  (INPUT)

CARD  1 = [HEADER
CARD  2 = [LEVER MECHANISM WITH REDUNDANT INERTIAS  -- 17
CARD  3 = [GRAPH
CARD  4 = [I 2 , C 3 , I 4 , R 5 , 1 1 2 6 , TF 6 7 ,
CARD  5 = [SE 1 , 1 7 4 3 5 .
CARD  6 = [GREND
CARD  7 = [PARAMETERS
CARD  8 = [I2= 2.0  C3= 8.0  I4= 3.0  R5= 2.0
CARD  9 = [TF67= 2.5
CARD 10 = [PREND
CARD 11 = [SOURCES
CARD 12 = [E1= CONSTANT 10.0
CARD 13 = [SREND
CARD 14 = [SOLUTION
CARD 15 = [TIN= 0.0  TFIN= 7.5  DT= 0.1
CARD 16 = [ICS= 0.0 0.0 .
CARD 17 = [TPLOT 0.0 7.5 1 P2 Q3 .
CARD 18 = [SLEND
CARD 19 = [JOBEND

END-OF-FILE
```

Figure 1-10. ENPORT-4 job deck for the lever mechanism.

```
SYSTEM BOND GRAPH

NODE    BONDS
 I        2
 C        3
 I        4
 R        5
 1        1  2  6
 TF       6  7
 SE       1
 1        7  4  3  5

GRAPH ANALYSIS.

 THE GRAPH HAS    8 NODES AND    7 BONDS.

 POWER FLOWS
BOND  FROM TO
  1   SE → 1
  2   1  → I
  3   1  → C
  4   1  → I
  5   1  → R
  6   1  → TF
  7   TF → 1

THE NODE-BOND INCIDENCE MATRIX WITH CAUSALITY
+ = STROKE END,  - = NON STROKE END.
1  I       2   0  0   0   0
2  C      -3   0  0   0   0
3  I      -4   0  0   0   0
4  R      -5   0  0   0   0
5  1       1  -2  6   0   0
6  TF     -6   7  0   0   0
7  SE     -1   0  0   0   0
8  1      -7   4  3   5   0
THERE ARE   5 FIELD BONDS AND   2 JUNCTION BONDS.

THE BONDS INTERNAL TO THE JUNCTION STRUCTURE ARE --
  7  6

THE  2 INDEPENDENT ENERGY VARIABLES ARE --
P 2 Q 3

THE  1 DEPENDENT ENERGY VARIABLES ARE --
P 4

THE  1 DISSIPATION VARIABLES ARE --
E 5

THE 1 SOURCE VARIABLES ARE --
E 1
```

Figure 1-11. ENPORT-4 analysis and set-up of the lever mechanism model.

```
THE A MATRIX

 -4.839+00    -3.871E+00
  5.000E+00    0.

THE B MATRIX

 1.935E-01
 0.

THE E MATRIX

 0.
 0.

THE EIGENVALUES OF THE A MATRIX...

NUMBER     REAL PART          IMAGINARY PART    TIME CONSTANT

   1       -2.4194E+00          3.6744E+00        -4.1333E-01
   2       -2.4194E+00         -3.6744E+00        -4.1333E-01



THE STEADY STATE X-VECTOR IS...
  .3664E-13   .4999E+00
```

Figure 1-12. ENPORT-4 analysis and formulation of the lever mechanism system.

indicate that P2 approaches zero (no mass motion), while the spring deflects 0.5 units (Q3 = 0.4999...). The eigenvalues suggest the possibility of a damped oscillation.

Key variable results are shown in Figure 1-13, and indicate that the momentum (P2) approaches 0.0 in an oscillatory manner, while the spring deflection (Q3) similarly moves toward the value 0.5.

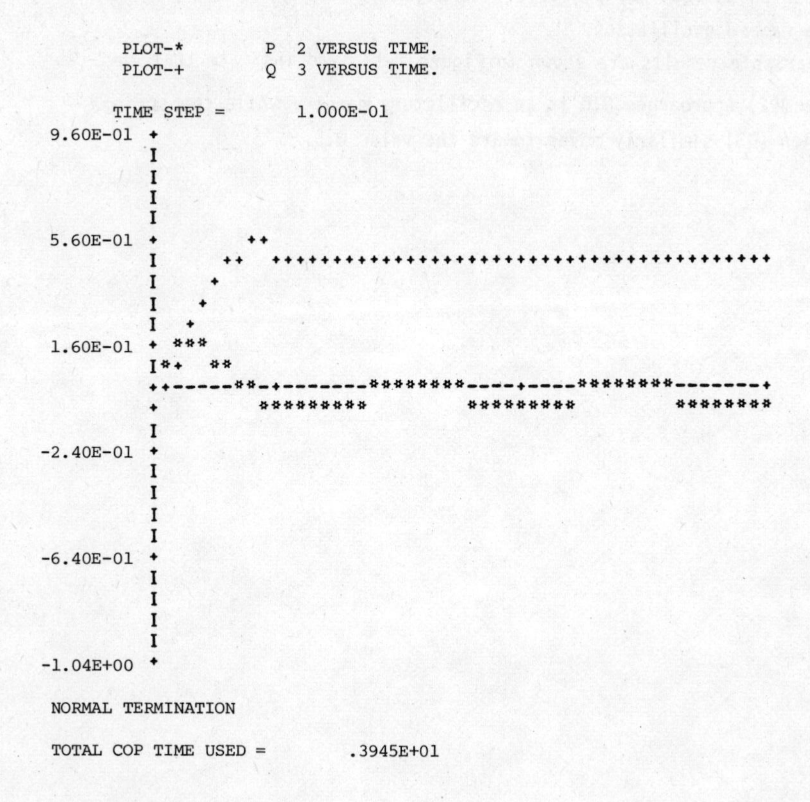

Figure 1-13. Plot of the dynamic response of key variables for the lever mechanism system.

1.3.4. Position control system

The purpose of this example is to suggest ways to handle device and component models in engineering systems, and to illustrate the feature of bond activation for signal coupling. Bond activation is useful in representing feedback control functions and active elements (e.g., transistors, instruments) in dynamic systems.

Figure 1-14 shows a radar pedestal unit with a drive motor and position control feedback. Details on the system and its bond graph model may be found in reference [9]. A device level multiport model is given in Figure 1-15. Note that some bonds are electrical (including voltages representing pure signals), while some are mechanical rotation. In Figure 1-16 a detailed bond graph model for each device is shown. The motor is field-controlled, and the armature current is assumed constant; hence the armature port effect has been absorbed into the motor model (GY56). Included in the system model represented in Figure 1-16 are field circuit inductance and resistance and mechanical inertia and resistance in the motor (I_f, R_f, I_m and R_m, respectively). The shaft is assumed to have compliance (C 10). The gears are represented as ideal (TF 11 12). The pedestal unit is assumed to have inertia and resistance (I_p and R_p), while the tachometer is modeled as an ideal instrument, converting angular velocity (ω_{15}) to voltage (e_{16}) by pure GY action. Bonds 15 and 16 are <u>active</u>, indicating that there is no load torque imposed on the pedestal by the tachometer, and that the output voltage from the tachometer (e_{16}) is independent of the current drawn (i_{16}).

The feedback control scheme is direct. The tach voltage is integrated (I_c) and scaled (GY1819) to arrive at a feedback voltage (e_θ). This voltage is compared to the command voltage (e_c), and the difference is used to drive the field port of the motor. And around the loop we have gone. By the use of storage elements (I,C), scaling (GY, TF), summation (0, 1), and bond activation, the general case of linear (state variable) feedback can be represented in terms compatible with the basic physical system model.

The job deck for ENPORT-4 is shown in Figure 1-17. The graph data is given by cards 3 to 11, where card 4 represents the command signal and summer, card 5 the motor, card 6 the shaft and gears, card 7 the

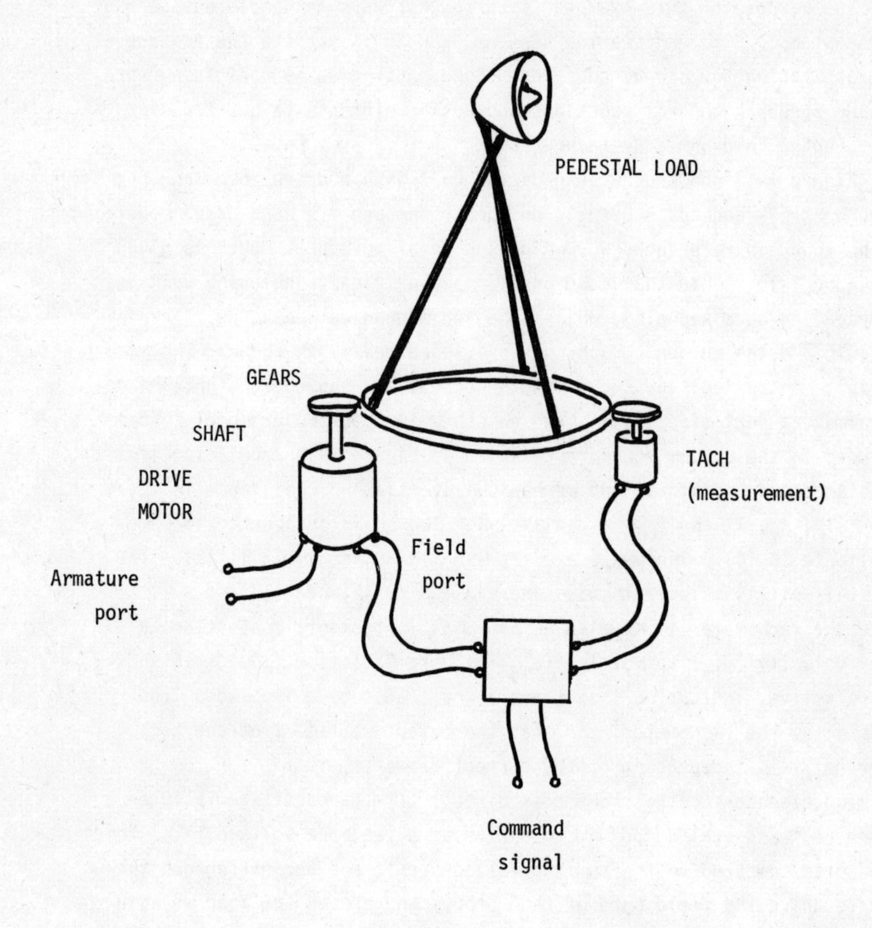

Figure 1-14. Radar pedestal position control system.

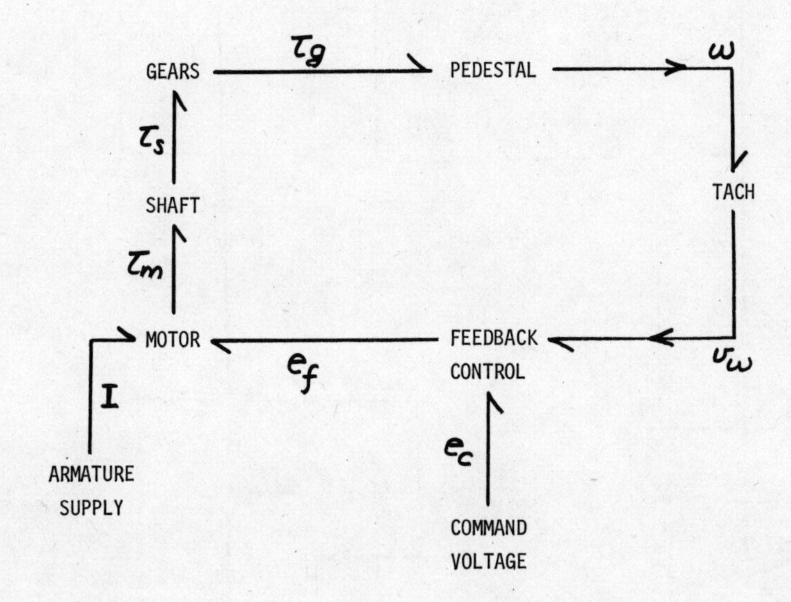

Figure 1-15. Device-level multiport model for the radar pedestal control system.

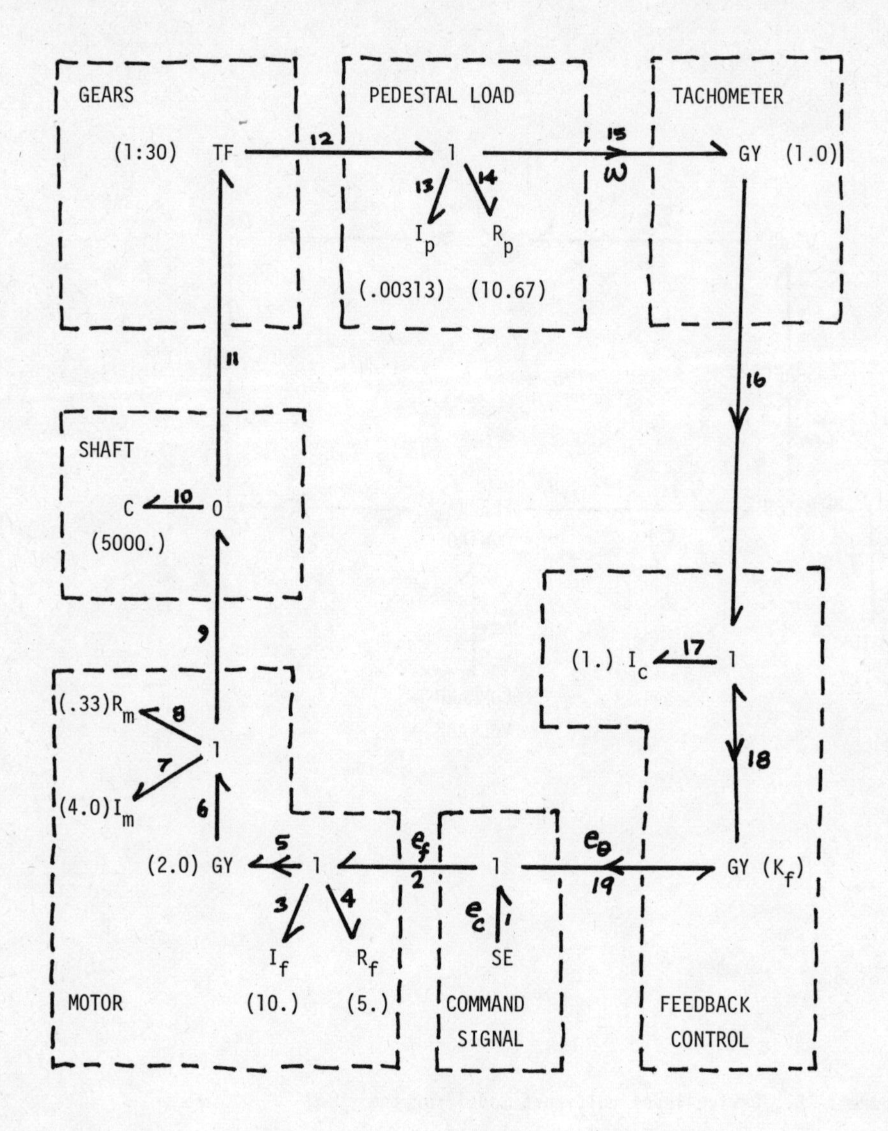

Figure 1-16. Bond graph model of the pedestal control system.

```
MSU

M S U   6 5 0 0   ENPORT              VERSION 4.1          UPDATE 50772
--------------------------------------------------------------------------------


CONTENTS OF TAPE2 (INPUT)

CARD  1 =  {HEADER
CARD  2 =  {RADAR PEDESTAL POSITION CONTROL -- RUN 5.
CARD  3 =  {GRAPH
CARD  4 =  {SE 1 , 1 1 2 1 9 ,
CARD  5 =  {1 2 3 4 5 , I 3 , R 4 , GY 5 6 , 1 6 7 8 9 , I 7 , R 8 ,
CARD  6 =  {0 9 10 11 , C 10 , TF 11 12 ,
CARD  7 =  {1 12 13 14 15 , I 13 , R 14 ,
CARD  8 =  {GY 15 16 , 1 16 17 18 , I 17 , GY 18 19 .
CARD  9 =  {ASSIGN 18 GY 1 .
CARD 10 =  {ACTIVATE 5 E 15 E 16 F 18 E 19 F.
CARD 11 =  {GREND
CARD 12 =  {PARAMETERS
CARD 13 =  {I3= 10.  R4= 5.0  GY56= 20.0  I7= 4.0  R8=0.333
CARD 14 =  {C10= 5000.   TF1112=   30.0
CARD 15 =  {I13= 0.00313   R14=   10.67
CARD 16 =  {GY1516= 1.0  I17= 1.0  GY1819= 1.0
CARD 17 =  {PREND
CARD 18 =  {SOURCES
CARD 19 =  {E1= CONSTANT 0.5
CARD 20 =  {SREND
CARD 21 =  {VARIABLES
CARD 22 =  {F15 E2 E6 F9 E12 E19
CARD 23 =  {VREND
CARD 24 =  {SOLUTION
CARD 25 =  {TIN=  0.0   TFIN=  0.1   DT==  0.0001
CARD 26 =  {ICS=0.0 .
CARD 27 =  {TPLOT  0.0  0.1  20  E2 .
CARD 28 =  {TPLOT  0.0  0.1  20  E6 .
CARD 29 =  {TPLOT  0.0  0.1  20  E12 .
CARD 30 =  {TPLOT  0.0  0.1  20  F15 .
CARD 31 =  {TPLOT  0.0  0.1  20  E19 .
CARD 32 =  {SLEND
CARD 33 =  {JOBEND
```

Figure 1-17. ENPORT-4 job deck for the radar pedestal example.

pedestal, and card 8 the tachometer and feedback control function. To ensure negative feedback the power direction was set as shown by card 9. The signal portion of the system is developed by activating selected bonds; for example, card 10 shows that on bond 15 the effort (torque) is to be suppressed. This is an explicit statement that the back torque from the tachometer on the radar pedestal may be assumed zero (a reasonable assumption for a good instrument).

Cards 12 to 17 specify the element parameters, which are grouped by device in the same manner as the graph data. This is a convenience for the user, not a necessity required by the program. The command signal (voltage corresponding to an angular position of the pedestal) is given as a constant of value 0.5 by cards 18 to 20.

In addition to the state (i.e., independent energy)variables and source variables, we wish to observe the behavior of a number of intermediate, or device-coupling, variables. These are specified by cards 21 to 23. Finally, the solution phase is controlled by cards 24 to 32, and includes zero initial conditions (card 26) and five plots, one for each output variable of interest. The variables could have been stacked on one plot, or otherwise grouped.

Results of processing are displayed beginning with Figure 1-18, where graph structure, power directions, and bond activation data are given. The key formulation variables are identified in Figure 1-19, and the associated A and B matrices are printed. The state vector, X(t), and source vector, U(t), are

$$X = \begin{bmatrix} P3 \\ P7 \\ Q10 \\ P13 \\ P17 \end{bmatrix} = \begin{bmatrix} \text{field coil flux linkage} \\ \text{rotor angular momentum} \\ \text{angle of twist of shaft} \\ \text{pedestal angular momentum} \\ \text{integral of feedback signal} \end{bmatrix}$$

and

$$U = [E1] = [\text{command signal (voltage)}].$$

The A and B matrices relate X and U by

$$\dot{X}(t) = A*X + B*U.$$

For example, from row five of the implied matrix equation we see that

```
SYSTEM BOND GRAPH

NODE    BONDS
 SE         1
 1          1    3   19
 1          2    3    4    5
 I          3
 R          4
 GY         5    6
 I          6    7    8    9
 I          7
 R          8
 O          9   10   11
 C         10
 TF        11   12
 1         12   13
 I         13        14   15
 R         14
 GY        15   16
 1         16   17   18
 I         17
 GY        18   19
 GRAPH ANALYSIS.
  THE GRAPH HAS    19 NODES AND    19 BONDS.
 POWER ON BOND 18 PRESET AS DESIRED
  POWER FLOWS
 BOND  FROM  TO
   1   SE →  1
   2   1  →  1
   3   1  →  1
   4   1  →  R
   5   1  →  GY
   6   GY →  1
   7   1  →  I
   8   1  →  R
   9   1  →  O
  10   O  →  C
  11   O  →  TF
  12   TF →  1
  13   1  →  I
  14   1  →  R
  15   1  →  GY
  16   GY →  1
  17   1  →  I
  18   GY →  1
  19   1  →  GY
 BOND  5 ACTIVATED E IS SUPPRESSED
 BOND 15 ACTIVATED E IS SUPPRESSED
 BOND 16 ACTIVATED F IS SUPPRESSED
 BOND 18 ACTIVATED E IS SUPPRESSED
 BOND 19 ACTIVATED F IS SUPPRESSED
```

Figure 1-18. ENPORT-4 analysis and set-up of the pedestal system.

```
THERE ARE    9 FIELD BONDS AND   10 JUNCTION BONDS.
THE BONDS INTERNAL TO THE JUNCTION STRUCTURE ARE --
 19 18 16 15 12 11  9 6 5 2
THE 5 INDEPENDENT ENERGY VARIABLES ARE --
P 3  P 7  Q10  P13 P17
THE 3 DISSIPATION VARIABLES ARE --
E 4 E 8 E14
The 1 SOURCE VARIABLES ARE --
E1

THE A MATRIX
 -5.000E=01  0.            0.          0.            -1.000E+00
  2.000E+02  -1.332E+00 -5.000E+03  0.             0.
  0.          4.000+00    0.         -1.043E-04     0.
  0.          0.          1.666E+02  -3.340E-02    0.

  0.          0.          0.          3.130E-03     0.
THE B MATRIX
  1.000E+00
  0.
  0.
  0.
  0.
THE EIGENVALUES OF THE A MATRIX...
NUMBER      REAL PART      IMAGINARY PART    TIME CONSTANT
   1        -6.6600E-01       1.4142E+02        -1.5015E+00
   2        -6.6600E-01      -1.4142E+02        -1.5015E+00
   3        -5.0000E+01      0.                 -2.0000E-02
   4        -1.6695E-02       1.1772E-02        -5.9898E+01
   5        -1.6695E-02      -1.1772E-02        -5.9898E+01
THE 6 OUTPUT VARIABLES ARE --
E19   F15   E12   F 9   E 6   E 2
THE C MATRIX.
 0.           0.           0.           0.            1.0000E+00
 0.           0.           0.           3.1300E-03   0.
 0.           0.           1.6665E+02  0.            0.
 0.           4.0000E+00   0.           0.            0.
 2.0000E+02   0.           0.           0.            0.
 0.           0.           0.           0.           -1.0000E+00
THE D MATRIX.
 0.
 0.
 0.
 0.
 0.
 1.0000E+00
```

Figure 1-19. ENPORT-4 analysis and formulation of the pedestal system.

$$\dot{(P17)} = 0.00313*(P13),$$

or the rate of change of (P17) equals the angular velocity of the pedestal (F15). Thus (P17) is the time integral of (F15), or the pedestal angular position.

The eigenvalues of A also are given in Figure 1-19, and suggest a pair of damped oscillatory responses and a stable system overall. The output vector, Y, is shown to be

$$Y(t) = \begin{bmatrix} E19 \\ F15 \\ E12 \\ F9 \\ E6 \\ E2 \end{bmatrix} = \begin{bmatrix} \text{feedback signal} \\ \text{pedestal angular velocity} \\ \text{gear torque on pedestal} \\ \text{motor angular velocity} \\ \text{motor-converted torque} \\ \text{field voltage (error signal)} \end{bmatrix}$$

This vector is related to X(t) and U(t) by

$$Y(t) = C * X + D * U,$$

where C and D are specified at the bottom of Figure 1-19.

Now the output of formulation data is complete and we can inspect some results. Figure 1-20 shows the solution conditions, and, since there are only constant sources, prints the steady-state conditions for both X and Y. The first entry in Y corresponds to E19, the voltage feedback signal corresponding to angular position of the pedestal; it has the value 0.5, as it should for position control with zero steady-state error. The other output variables all approach zero, indicating no motions or transmitted torques in the long run.

The five output variables have been plotted in Figures 1-21 (E2 and E6), 1-22 (E12 and F15), and 1-23 (E19). By reading through them in order, we see that the field voltage jumps to 0.5 immediately and stays (near) there, the motor-converted torque (E6) shows a lag, the gear torque delivered to the pedestal is increasing in an oscillatory fashion (E12), the pedestal angular velocity (F15) is starting to increase, and finally, the feedback signal (E19) is slowly increasing. The time range for all these plots is 0.002.

The next step in studying the system might be to recognize that the

```
T-INITIAL   =      0.                     T-FINAL  =          .1000E+00


DELTA-T  =        .1000E-03


STATE VARIABLE INITIAL CONDITIONS
XO( 1) =   0.
XO( 2) =   0.
XO( 3) =   0.
XO( 4) =   0.
XO( 5) =   0.
THE STEADY STATE X-VECTOR IS...
  .5689E-10  .2960E-12   .2276E-11   .1135E-07   .5000E+00
THE STEADY STATE Y-VECTOR IS...
  .5000E+00  .3553E-10   .3791E-09   .1184E-11   .1138E-07

  .2845E-08


SYSTEM VARIABLE BOUNDS
VARIABLE     MAXIMUM    MINIMUM
X( 1) =     9.9326E-03  0.
X( 2) =     6.3534E-03 -4.3977E-03
X( 3) =     5.1879E-04  0.
X( 4) =     5.2795E-03  0.
X( 5) =     6.9957E-07  0.
Y( 1) =     6.9957E-07  0.
Y( 2) =     1.6525E-05  0.
Y( 3) =     8.6431E-02  0.
Y( 4) =     2.5414E-02 -1.7591E-02
Y( 5) =     1.9865E+00  0.
Y( 6) =     5.0000E-01  5.0000E-01
U( 1) =     5.0000E-01  5.0000E-01
```

Figure 1-20. Partial results for the pedestal system.

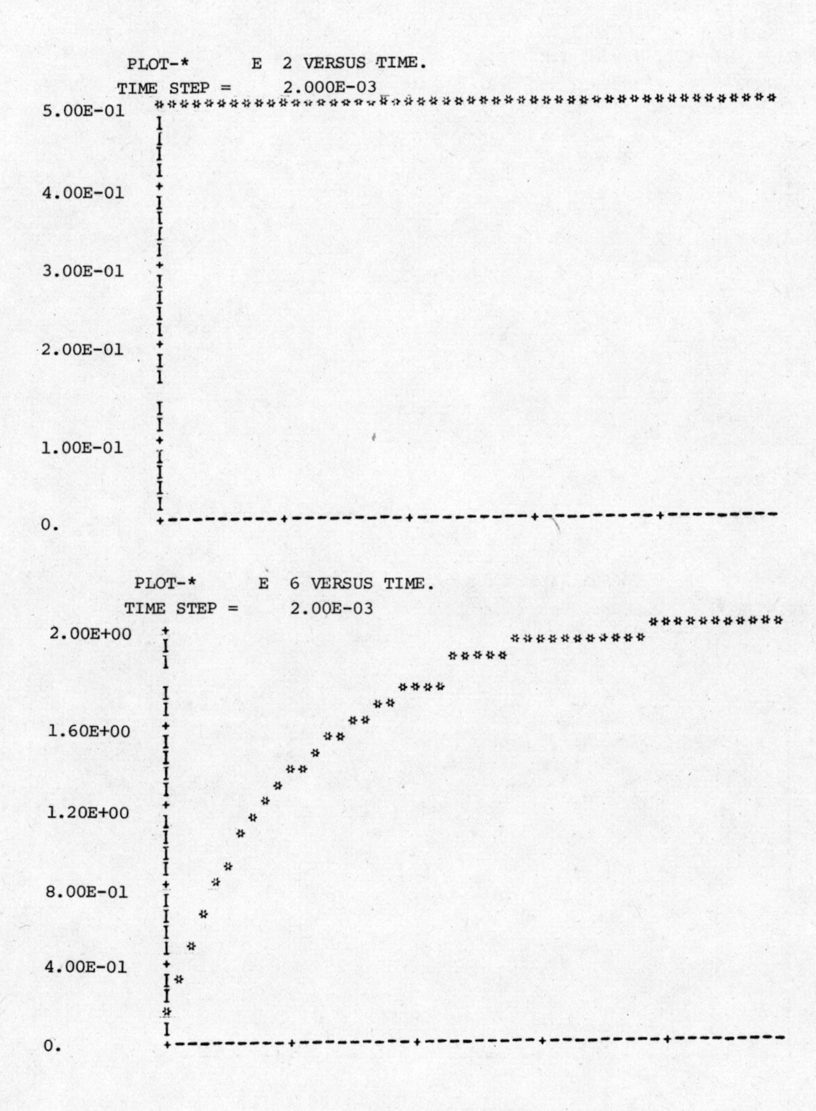

Figure 1-21. Dynamic response of the pedestal system - field voltage and converted torque.

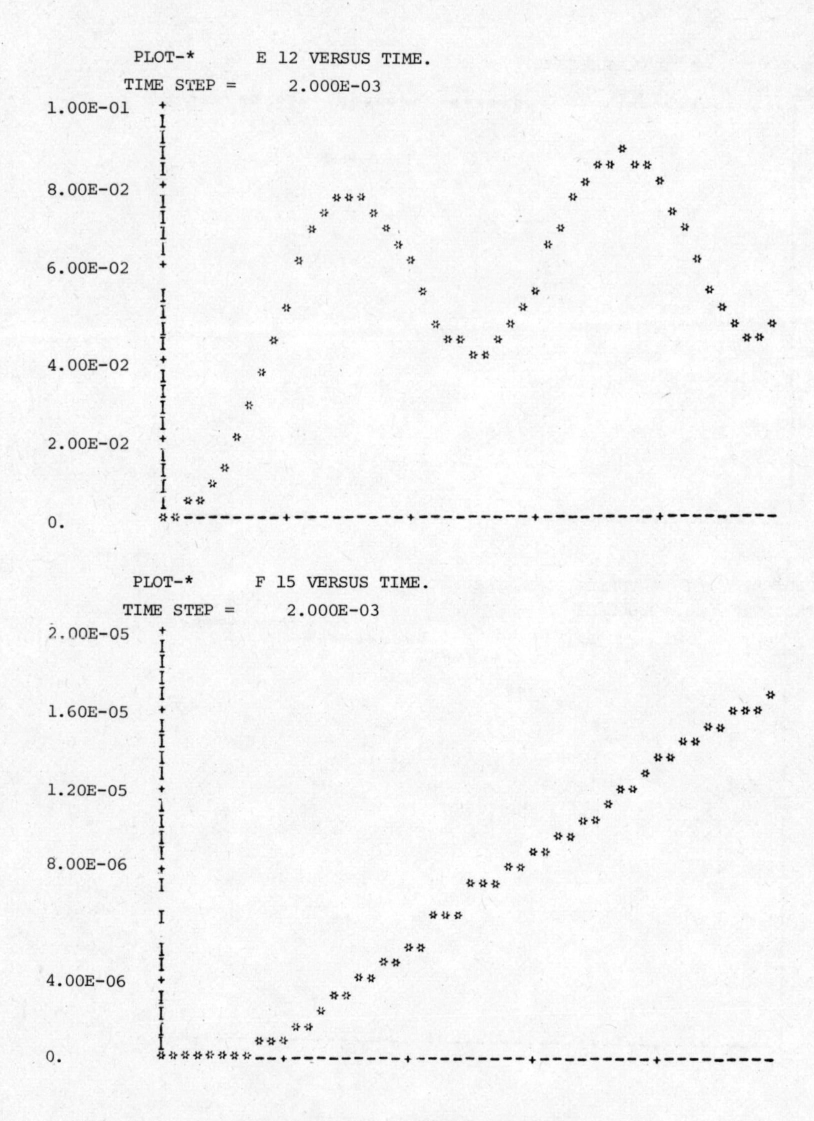

Figure 1-22. Dynamic response of the pedestal system - pedestal torque and angular velocity.

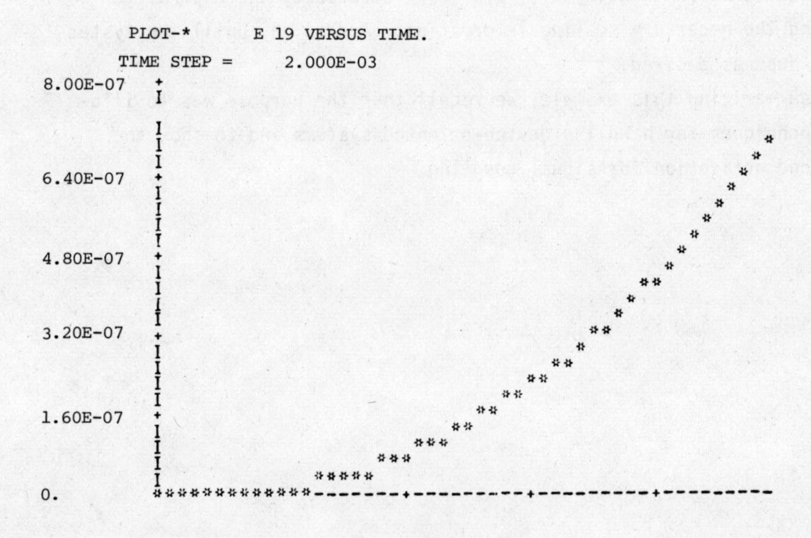

Figure 1-23. Dynamic response of the pedestal system - position feedback voltage.

speed of response of the motor field circuit suggests the elimination of the inductance effect (I3), thereby simplifying the system somewhat. Now a longer time span could be investigated efficiently and design modific-ations could be introduced in the feedback function. Also, the feedback function could be replaced by a simple (0-C) structure, the C parameter containing the necessary scaling information. And so on, until the system performs just as desired.

In summarizing this example, we recall that the purpose was to illu-strate techniques for handling device-oriented systems and to show the use of bond activation for signal coupling.

1.4 Parameter Definitions in ENPORT-4

As a convenience to users of ENPORT-4 we shall summarize in this section a standard set of parameter definitions expected by the program. Further discussion is found in Chapter 2.

1.4.1 Generalized variables.

<u>Power</u> E = effort
F = flow

<u>Energy</u> P = momentum (generalized)
Q = displacement (generalized)

Bond Graph Element	ENPORT-4 Definition	Parameter Units
R, resistance	$E = N_R * F$	$[N_R] = \frac{[E]}{[F]}$
C, capacitance	$E = N_R * Q$	$[N_C] = \frac{[E]}{[Q]}$
I, inertance	$F = N_I * P$	$[N_I] = \frac{[F]}{[P]}$

where N represents a number.

1.4.2 Mechanical translation variables.

Power E = force (F)

F = velocity (V)

Energy P = momentum (p)

Q = displacement (x)

Bond Graph Element	ENPORT-4 Definition	Parameter Units
R, dissipation	$F = N_R * V$	$[N_R] = \frac{[F]}{[V]}$
C, compliance	$F = N_C * x$	$[N_C] = \frac{[F]}{[x]}$
I, inertance	$V = N_I * p$	$[N_I] = \frac{[V]}{[p]}$

NOTE that N_C is equivalent to Stiffness, and N_I is equivalent to inverse inertia, as defined in common mechanical engineering usage.

1.4.3 Mechanical rotation variables.

<u>Power</u> E = torque (I)

F = angular velocity (ω)

<u>Energy</u> P = angular momentum (h)

Q = angular position (θ)

Bond Graph Element	ENPORT-4 Definition	Parameter Units
R, dissipation	$\tau = N_R * \omega$	$[N_R] = \frac{[\tau]}{[\omega]}$
C, compliance	$\tau = N_C * \theta$	$[N_C] = \frac{[\tau]}{[\theta]}$
I, inertance	$\omega = N_I * h$	$[N_I] = \frac{[\omega]}{[h]}$

NOTE that N_C is equivalent to torsional stiffness, and N_I is equivalent to <u>inverse</u> moment of inertia, as defined in common mechanical engineering usage.

1.4.4 Hydraulic variables.

Power E = pressure (P)
F = volume flow (Q)

Energy P = pressure-momentum (Γ)
Q = volume (V)

Bond Graph Element	ENPORT-4 Definition	Parameter Units
R, dissipation	$P = N_R * Q$	$[N_R] = \frac{[P]}{[Q]}$
C, capacitance	$P = N_C * V$	$[N_C] = \frac{[P]}{[V]}$
I, inertance	$Q = N_I * \Gamma$	$[N_I] = \frac{[\Gamma]}{[Q]}$

NOTE that N_C is equivalent to inverse fluid compliance or tank capacitance, and N_I is equivalent to inverse fluid inertia, as defined in common hydraulic engineering usage.

1.4.5 Electrical network variables.

<u>Power</u> E = voltage (v)
F = current (i)

<u>Energy</u> P = flux linkage (λ)
Q = charge (q)

Bond Graph Element	ENPORT-4 Definition	Parameter Units
R, resistance	$v = N_R * I$	$[N_R] = \frac{[v]}{[i]}$
C, capacitance	$v = N_C * q$	$[N_C] = \frac{[v]}{[q]}$
I, inductance	$i = N_I * \lambda$	$[N_I] = \frac{[i]}{[\lambda]}$

NOTE that N_C is equivalent to <u>inverse</u> capacitance, and N_I is equivalent to <u>inverse</u> inductance, as defined in common electrical engineering usage.

1.4.6 Thermal conduction variables.
(quasi-bonds)

<u>Power</u> E = temperature (T)

F = heat flow (q)

<u>Energy</u> P = (not used)

Q = thermal energy (u)

Bond Graph Element	ENPORT-4 Definition	Parameter Units
R, resistance	$T = N_R * q$	$[N_R] = \frac{[T]}{[q]}$
C, capacitance	$T = N_C * u$	$[N_C] = \frac{[T]}{[u]}$
I, inertance	----	----

NOTE that N_C is equivalent to the <u>inverse</u> of thermal capacitance, as defined in common thermal engineering usage.

1.4.7 Transformer and gyrator parameters.

The transformer (TF) and gyrator (GY) elements each have a single parameter, or modulus. These are defined in generalized bond graph terms by ENPORT-4 as follows:

<u>gyrator</u> $\xrightarrow{1}$ GY $\xrightarrow{2}$ (modulus R)

$$\begin{cases} E2 = R * F1 \\ E1 = R * F2 \end{cases}$$

The modulus is "R", where $[R] = \frac{[E]}{[F]}$.

<u>transformer</u> $\xrightarrow{1}$ TF $\xrightarrow{2}$ (modulus N)

$$E1 = N * E2$$
$$F2 = N * F1$$

where N is a pure number (dimensionless). For the definition given above, <u>the line code description must be TF 1 2</u>. (That is, bond 1 is the first named; bond 2 is the second named.)

One remark is in order when either GY or TF is used as a transducer element. It is not generally possible to define the effort-flow pair on each side of the 2-port independently, use a single modulus and have power in equal to power out numerically. A suggested modeling technique for handling this case is illustrated below for electromechanical conversion.

$\xrightarrow{1}$ GY $\xrightarrow{2}$ $\equiv$ $\vdash\xrightarrow[i_1]{v_1}$ GY $\xrightarrow[\omega_2]{\tau_2}\dashv$

To meet the conditions

$$\tau_2 = K_m * i_1,$$
$$v_1 = K_s * \omega_2$$

where $v_1\, i_1 = (\frac{K_s}{K_m}) * \tau_2\omega_2$ is a true physical power condition, then use

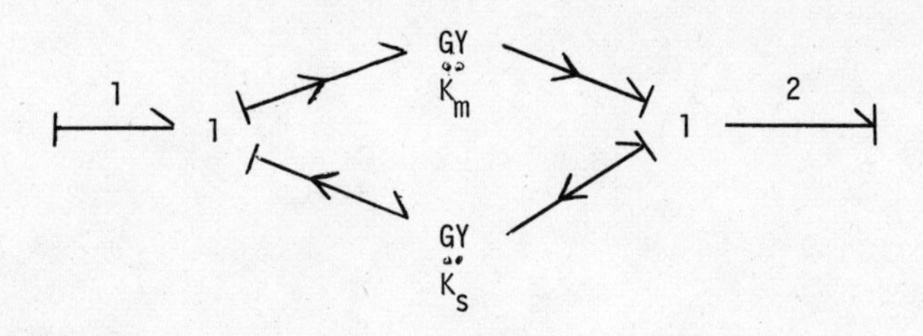

NOTE that, as far as numerical results in ENPORT-4 are concerned, power on bond 1 will not be equal to power on bond 2 when the true physical powers are equal.

1.5 References

This section lists a number of bond graph-related references. For the inexperienced reader the two likely to be of most immediate assistance are System Dynamics [2], which gives considerable detail about bond graph methodology, and the Special Collection on Bond Graph Modeling of the Trans. A.S.M.E. [7], which gives an excellent perspective on the (1972) state-of-the-art and includes as well an extensive bibliography.

There are several journals in which bond graph articles have appeared somewhat regularly, including Trans. A.S.M.E., Journal of Dynamic Systems, Measurement, and Control; Journal of the Franklin Institute; the Joint Automatic Control Conference (JACC) Proceedings; and the International Federation of Automatic Control (IFAC) Proceedings.

For those readers who prefer a succinct introduction to bond graphs in German, there is Dr. J. Thoma's little gem [12], and for those who are interested in a historical view there is Professor Henry Paynter's fascinating book [13], which contains the seeds of many of the flowers currently in bloom.

[1] Karnopp, D.C. and R.C. Rosenberg, Analysis and Simulation of Multiport Systems (M.I.T. Press, Cambridge, Mass., 1968).

[2] Karnopp, D.C. and R.C. Rosenberg, System Dynamics: A Unified Approach, Wiley-Interscience, N.Y., N.Y., 1975.

[3] Rosenberg, R.C. and D.C. Karnopp, "A definition of the bond graph language," Trans. ASME, J. Dynamic Systems, Measurement, and Control, (September, 1972).

[4] Paynter, H.M. and D.C. Karnopp, "Design and control of multiport engineering systems," Proc. IFAC Tokyo Symp. on Systems Engineering for Control System Design, 443-454 (1965).

[5] Karnopp, D.C., "Power-conserving transformations: physical interpretations and applications using bond graphs," J. of the Franklin Institute 288, 3, 175-201 (September, 1969).

[6] Karnopp, D.C. and R.C. Rosenberg, "Application of bond graph techniques to the study of vehicle drive line dynamics," Trans. ASME, J. Basic Engineering, 355-362 (June, 1970).

[7] "Special collection on modeling applications of the bond graph language," Trans. ASME, J. of Dynamic Systems, Measurement, and Control, (September, 1972).

[8] Keonig, H.E., et al., Analysis of Discrete Physical Systems (Mc-Graw-Hill, Inc., New York, 1967).

[9] Karnopp, D.C. and R.C. Rosenberg, "Power Bond Graphs: A New Control Language," Control Engineering, May, 1968, 79-82.

[10] Rosenberg, R.C., "State Space Formulation for Bond Graph Models of Multiport Systems," Trans. ASME, J. Dynamic Sys., Measurement, and Control, 34-40, March, 1971.

[11] Rosenberg, R.C., "Modeling and Simulation of Large-Scale, Linear Multiport Systems," AUTOMATICA, Vol. 9, 87-95 (1973).

[12] Thoma, J.V., Grundlagenund Anwendungen der Bonddiagramme, Verlag W. Girardet, Essen, W. Germany, 1974.

[13] Paynter, H.M., Analysis and Design of Engineering Systems, M.I.T. Press, Cambridge, Mass., 1960.

1.6 Quirk Sheet and Remarks

This section contains several important pages which can be used to exchange information about the ENPORT-4 program.

1.6.1 Current quirks

In ENPORT-4.2 (i.e., version 2 of ENPORT-4), there are no known quirks.

1.6.2 User Information Update Sheet

To keep yourself informed of ENPORT-related developments, use the User Information Update Sheet (or a suitable copy). This sheet is enclosed as Appendix B.

1.6.3 User Bug and Usage Sheet

This page speaks for itself. Read it and use it as indicated. It is enclosed as Appendix C.

CHAPTER 2. HOW TO USE ENPORT-4

ENPORT-4 operates in either of two modes -- batch or interactive. The batch mode is designed to allow flexible job construction and the interactive mode is naturally highly flexible, so the style of the two modes is similar in many respects.

Generally in batch mode the input is by cards and the output is by printer, whereas typically both input and output are by teletype in interactive mode.

2.1 Batch Mode

2.1.1 Sample job description

A sample input deck for an ENPORT-4 job in batch mode is shown in Figure 2-1. This job is single-pass, in that it progresses from one basic operation to the next until a complete run has been made, and then it stops. By studying this type of job first an understanding of the basic operations and their sequencing can be gained. Then the full flexibility of ENPORT-4 can be examined by reading Part 2.2 on batch job structure.

In Figure 2-1 the actual card images start to the right of the left square bracket on each labeled line. These cards are the data for the ENPORT program. Control cards for calling ENPORT will vary from computer to computer (or operating system to operating system), and are not shown here. Section 2.4 lists the control cards for the CDC 6500 computer at Michigan State University.

The first operation is to specify the desired system outputs during ENPORT processing by use of the DISPLAY command.

The next operation is to label the problem. HEADER does that.

The next operation is to read in a definition of the bond graph, and to assign powers and causality to it. GRAPH does that. At this point state variables are selected.

The next operation is to read in values of parameters for all C, I, R, TF and GY elements. PARAMETERS does that. At this point the system state equations are formulated, and the A, B, and E matrices can be printed out, if desired.

The next operation is to read in definitions of all source variables. SOURCES does this.

Next output variables (other than state and source variables) must be specified. VARIABLES does this.

At this point a solution may be calculated and output data requested. SOLUTION reads in time control data, initial conditions, and output requests. Plots and prints of responses are generated.

Finally, JOBEND terminates this run of ENPORT.

A convenient summary of the major control words is given in Table 2-1.

```
M S U   6 5 0 0   ENPORT            VERSION 4.2           UPDATE
--------------------------------------------------------------

CONTENTS OF TAPE7

          INFORMATION FOR PROCESSING
          ----------------------------------------------------

CARD  1 = [HEADER
CARD  2 = [SAMPLE ENPORT JOB - BASIC SEQUENCE  (2.1)
CARD  3 = [DISPLAY ANALYSIS CLASS STATE OUTPUT .
CARD  4 = [GRAPH
CARD  5 = [SE 1 ,  C 2 ,  TF 3 4 ,  I 5 ,  R 6 ,  SF 7 ,
CARD  6 = [1 1 2 3 ,  0 4 5 6 7 .
CARD  7 = [GREND
CARD  8 = [PARAMETERS
CARD  9 = [C2= 4.0  TF34= 3.0  I5= 2.0  R6= 0.2
CARD 10 = [PREND
CARD 11 = [SOURCES
CARD 12 = [F7= STEP 10.0  0.5
CARD 13 = [E1= CONSTANT 12.5
CARD 14 = [SREND
CARD 15 = [VARIABLES
CARD 16 = [E2 F5 E6 F6 E7 F7
CARD 17 = [VREND
CARD 18 = [SOLUTION
CARD 19 = [TIN= 0.0  TFIN= 10.0  DT= 0.1
CARD 20 = [ICS=  0.0 0.0 .
CARD 21 = [BPLOT 0.0 10.0 0.1 X1 U1 .
CARD 22 = [BPLOT 0.0 10.0 0.1 PW6 PW7 .
CARD 23 = [BPLOT 0.0 10.0 0.1 E2 Q2 F5 P5 .
CARD 24 = [BPRINT 0.0 10.0 0.1 X1 X2 E2 F5 .
CARD 25 = [SLEND
CARD 26 = [JOBEND
```

Figure 2-1. Sample job deck for batch mode - single pass.

cards	initial control word	final control word	function	section
3	DISPLAY	none	control of information displayed	2.1.7
1- 2	HEADER	none	one line of label for the problem	---
4- 7	GRAPH	GREND	line code for the bond graph	2.1.2
8- 10	PARAMETERS	PREND	specification of parameters	2.1.3
11-14	SOURCES	SREND	specification of source functions	2.1.4
15-17	VARIABLES	VREND	specification of output variables	2.1.5
18-25	SOLUTION	SLEND	control data for solution and output	2.1.6
26	JOBEND	none	terminates this job	---

Table 2-1. Basic operation sequence in ENPORT-4.

(Refer to Figure 2-1.)

2.1.2. Graph processing

GRAPH is used to define the bond graph structure and to control assignment of power directions, causality and activation. The graph processing section must be terminated by GREND.

The bond graph is specified by giving a list of node-bond groups, called the line code. A node-bond group is constructed by first listing a multiport element (e.g., 0, R, TF, etc.), followed by all its bonds. The order of bonds within a group is unimportant, except in the case of TF elements (see Section 2.1.3 on parameter specification). There are as many node-bond groups as there are elements in the graph.

The order of node-bond groups is basically arbitrary. However, it is possible to influence the direction of powers by the order of the groups (see ASSIGN below). Each group is demarcated by a comma, and the last group is ended by a period. At least one space must separate each specific item in the line code, including commas. Bonds must be numbered consecutively, starting from 1. The largest bond number serves as the bond count.

Within the graph processing section, three options exist. ASSIGN allows powers to be directed by the user. CAUSAL allows the user to select causality when the choice is optional. ACTIVATE allows the user to activate selected bonds.

Power directions.

ASSIGN -- Power directions are chosen by ENPORT-4 according to the following rules: the program will

(1) direct powers into C, I and R ports;
(2) direct powers out of SE and SF ports;
(3) direct powers through TF and GY elements; and
(4) for all bonds still undirected, direct power from the first named element in the line code to the second-named element.

Thus the order of node-bond groups in the line code can be used to influence the assignment of powers, particularly on bonds between junctions.

Any bond(s) may have its (their) power(s) directed explicitly by using the ASSIGN option. Give the bond number, the 'power from' element, and the 'power to' element. The list consists of sets of such data, separated by spaces between items, and terminated by a period. An example of usage is given below.

Causality assignment.

CAUSAL -- Causality is assigned to the graph in accordance with standard bond graph practice. (See, for example, references [1,2]).

If an arbitrary causal choice can be made, the user may (but does not have to) specify the choice desired. Give the bond number, the element, and the input variable (E or F) to that element. Any (consistent) set of bonds may be so directed. The list items are separated by spaces, and the list is terminated by a period. An example of usage is given below.

Bond activation.

ACTIVATE -- In a causal graph, any set of bonds may be activated (i. e., made into pure signal couplers) by use of the ACTIVATE option. The list gives a bond followed by a suppressed signal (E or F) followed by a bond, and so on, terminated by a period. Items in the list are separated by spaces.

Any combination of the options ASSIGN, CAUSAL and ACTIVATE may be selected, but they must be used in that order.

Example of usage.

```
GRAPH
 SE 1, SE 2, 1 1 2 3, 1 3 4 5 6, R 4, C 5, 0 6 7 8, R 7 ,
 1 8 9 10 11, R 9, TF 10 12, TF 11 13, 0 12 13 14 15 ,
 R 14, 0 15 16 17, 1 16 18 19, SE 18, R 19, 1 17 20 21 ,
 R 20, C 21 .
 ASSIGN 14 R 0 .
 CAUSAL 7 R F  9 R E  20 R F .
 ACTIVATE 11 E  12 F .
 GREND
```

In this example, power will be assigned from R to 0 on bond 14.

Causality will be chosen so that on bond 7 input to R will be F(flow), on bond 9 input to R will be E(effort), and on bond 20 input to R will be F.

Activation will occur on bond 11 by E being suppressed and on bond 12 by F being suppressed.

2.1.3. Parameter specification

PARAMETERS is followed by a labeled list defining all R, C, I, TF and GY characteristics. The list must be terminated by PREND. Each element is identified by its type, followed by its bonds, followed by '=', without spaces. The bond order must be identical to that used in the line code.

The characteristic for each element is defined as follows:

CAPACITANCE - C i : $e_i = k*q_i$

C i j : $e_i = k_{ii}*q_i + k_{ij}*q_j$

$e_j = k_{ji}*q_i + k_{jj}*q_j$

where C i is a 1-port, C i j is a 2-port, and the generalization to n-ports should be clear. For multiports, the k's should be entered by rows (e. g., k_{ii}, k_{ij}, k_{ji}, k_{jj}). Note that efforts are always defined in terms of displacements.

INERTANCE - I i : $f_i = k*p_i$

I i j : $f_i = k_{ii}*p_i + k_{ij}*p_j$

$f_j = k_{ji}*p_i + k_{jj}*p_j$

where I i is a 1-port, I ij is a 2-port, and the generalization to n-ports should be clear. For multiports, the k's should be entered by rows. Note that flows are always defined in terms of momenta.

RESISTANCE - R i : $e_i = k*f_i$

R i j : $e_i = k_{ii}*f_i + k_{ij}*f_j$

$e_j = k_{ji}*f_i + k_{jj}*f_j$

Note that efforts are always defined in terms of flows.

TRANSFORMER - TF i j : $e_i = k*e_j$

$f_j = k*f_i$

where i is the first-named bond in the line code definition and j is the second-named bond. 'k' is the dimensionless modulus.

GYRATOR - GY i j : $e_i = k*f_j$

$$e_j = k*f_i$$

where 'k' is the resistive modulus.

Remarks on parameter specification.

(1) The definition of multiport characteristics is independent of the final causality assigned to the graph.

(2) Historical parameter definitions, such as stiffness, capacitance, and inertia must be related in each instance to the required 'k' for the element. For example, in mechanics the 'k' for an inertia is the reciprocal of the mass (see section 1.4).

Example of usage.

```
PARAMETERS
  C8= 2.5  R12=  0.75  TF23=  1.5
  I711=  1.0  0.2  0.2  4.0
PREND
```

The following element characteristics are implied:

(C8) $e_8 = 2.5 * q_8$

(R12) $e_{12} = 0.75 * f_{12}$

(TF23) $e_2 = 1.5 * e_3$

$f_3 = 1.5 * f_2$

(I711) $f_7 = 1.0 * p_7 + 0.2 * p_{11}$

$f_{11} = 0.2*p_7 + 4.0 * p_{11}$

2.1.4 Source definition

SOURCES is followed by a set of source variable definitions, one per line, taken from Table 2-2. The source list must be terminated by SREND. If there are no sources SOURCE, SREND may be omitted. In the source variable list, variables are referred to as Ei=, Fj=, or Uk=, where the first refers to the effort on bond i, the second to the flow on bond j, and the third to the kth element in the source vector, U_k.

Example of usage.

```
SOURCES
E7=  STEP  12.5 0.25
F2= SIN   2.0  366.  0.
SREND
```

Name	Arguments	Definition
CONSTANT	a	u(t)= a, all t.
SIN	a w ∅	u(t)= a*sin(w*t + ∅)
COS	a w ∅	u(t)= a*cos(w*t + ∅)
STEP	a T	u(t)= 0., $t \le T$; = a , $t \ge T$.
RAMP	s T	u(t)= 0., $t < T$. = s*t, $t \ge T$.
SQWAVE	a_1 a_2 T_0 T_1 T_2	(see figure SQW below)
TRWAVE	a_0 a_1 a_2 T_0 T_1 T_2	(see figure TRW below)
PULSE1	a T	u(t)= a*exp(-t/T)
PULSE2	a_1 a_2 T_1 T_2	u(t)= a_1*(1 - exp(-t/T_1)) + a_2*(1 - exp(-t/T_2))
TSERIES	n c_i, i= 0, 1, .. n ($n \le 10$)	$u(t)= c_0 + c_1t^1 + c_2t^2 + ..$
FSERIES	n a_1 b_1 a_2 b_2 ... ... a_n b_n w ($n \le 10$)	$u(t) = \sum_{i=1}^{n} (a_i \sin iwt + b_i \cos iwt)$
WSTATE	K_i, i = 1 to NX	$u(t) = \sum_{i=1}^{NX} K_iX_i$ (NX=no. of state vbls)

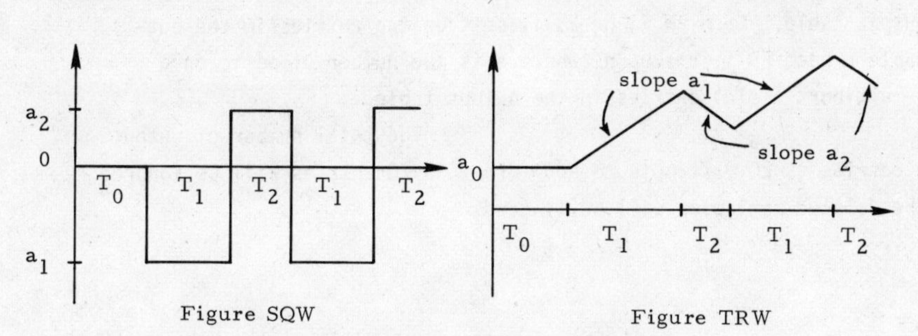

Figure SQW

Figure TRW

Table 2-2. Source definitions in ENPORT-4.

2.1.5 Output variable setup

VARIABLES is used to set up a table of system variables whose response is of interest. The variable list must by terminated by VREND. Reference may be made to any of the following variable types on any bond "i", where "i" = 1,2,3,,,NB (NB is the highest bond number used in the graph).

Ei , effort on bond i;
Fi , flow on bond i;
Qi , displacement on bond i;
Pi , momentum on bond i;
PWi , power on bond i;
ENi , net energy transferred on bond i;
ESj , energy stored in a C or I field to which bond j may be any one of the attached bonds.

Example of usage.

```
VARIABLES
Q1  P3  PW2  F7  EN4  ES5
VREND
```

The variables available for printing or plotting will be displacement on bond 1, momentum on bond 3, power on bond 2, flow on bond 7, energy transfer on bond 4, and energy stored in the C or I field to which bond 5 is attached. All state and source variables are automatically available for output, and need not be specified under this command, If no other variables than state and source variables are of interest, VARIABLES, VREND may be omitted. Each E i, F i, Q i, and P i sets up one entry in the output table. Each PW i, or EN i, sets up two entries in the output table. Each ES i sets up n (where n is the number of ports on a given multiport field) entries in the output table.

The total number of output table entries is 20; attempts to add more than 20 entries will be ignored and the deleted variables will be flagged.

2.1.6. Solution phase

SOLUTION initiates the solution phase, and must be terminated by SLEND. At least three cards are indicated, one for time control data, one for initial conditions, and one for output request.

Time control data.

TIN= a   TFIN= b   DT= c   NSTEPS= d   ACC= e

will cause a solution to be calculated from initial time 'a' to final time 'b' in time intervals of 'c'. The actual interval used for computation may be adjusted by the program if convergence and accuracy considerations require it; a warning is given. The original DT value is maintained for output printing and plotting.

The maximum number of steps to be calculated by the program can be set by NSTEPS. In the example above the number would be 'd'. The actual number of steps calculated depends upon the ratio (TFIN-TIN)/(actual computing interval). Typically, the effective number of steps is (TFIN-TIN)/DT.

Accuracy may be indicated by the use of ACC. In the case above ACC= 'e', which implies that the elements of the state vector will be accurate to a tolerance of 'e' for the case of no or constant forcing over the time range specified. In general, the tighter the accuracy, the smaller the actual computing interval used. Hence, the larger the number of steps required to cover the time range.

Default values.

Any or all of the values DT, NSTEPS, and ACC may be omitted, with the following default values used:

DT = 1/ (maximum magnitude element of the A matrix) ;

NSTEPS = 100,000 ;

ACC = 0.001 .

If the program is unable to meet the accuracy required in the maximum number of steps available using an appropriate value of time computing interval, a warning is given. This condition suggests either excessive accuracy requirements, a time range (TFIN-TIN) badly related to the system dynamics, or a system with a very wide spread of time response characteristics.

Initial conditions.

ICS= a b c d .

The first four state variables will be assigned the values 'a', 'b', 'c' and 'd', respectively, at time TIN. If insufficient data is provided, zero is used as the default value. Excess data is ignored. Data must be terminated by a period.

Plotting and printing.

Plotting and printing are initiated by the BPLOT, BPRINT, TPLOT, TPRINT, and PPLOT requests. A list of such requests may be used, one per card.

BPLOT t_1 t_2 dt vbl_1 vbl_2 . . . vbl_n .

or

BPRINT t_1 t_2 dt vbl_1 vbl_2 . . . vbl_n .

will cause plotting or printing from time 't_1' to time 't_2' in steps of 'dt'. 'dt' should be an integer multiple of DT. Output variables will be vbl_1, vbl_2, . . . , vbl_n. The list must end with a period, and spaces must separate the items. Up to five variables may be plotted on one graph, and up to five variables may be printed per request.

BPLOT is batch plot and BPRINT is batch print, both outputs sized for a line printer.

TPLOT is teletype plot and TPRINT is teletype print, both outputs sized for a teletype page. Actual sizes are

BPLOT = 8.5 x 11 plot versus time (dimensions in inches)
TPLOT = 6.5 x 5.5 plot versus time
PPLOT = 8.5 x 11 phase plot, vbl_1 versus vbl_2

The batch plot has a maximum of 101 time stages and the teletype plot has a maximum of 51 time stages. TPLOT may be called for in a batch job, as shown in the examples of section 1.3.

Allowable variable types are

E i , meaning effort on bond i;

F i ,   meaning flow on bond i;

Q i ,   meaning displacement on bond i;

P i ,   meaning momentum on bond i;

PW i ,   meaning power on bond i;

EN i ,   meaning net energy transferred on bond i, and

ES i ,   meaning energy stored in C- or I-field with port i.

Example of usage.

```
SOLUTION
TIN=  10.0  TFIN=  30.0   DT=  0.5
ICS=  7.5  0.0  3.5  0.0  .
BPLOT  10.0  30.0  0.5  E2  F7  Q8  PW9  .
BPLOT   24.0  30.0  1.0  EN9  .
BPRINT  10.0  30.0  2.5  X3  X4  U7  .
SLEND
```

The solution will be calculated from time 10.0 to time 30.0 in intervals of 0.5.

The initial conditions are X1(10.) = 7.5, X2(10.) = 0., X3(10.) = 3.5, and X4(10.) = 0.0.

There will be two plots and a printing. The first plot will go from 10. to 30. in time steps of 0.5 (1*0.5= 0.5), and the ordinate variables will be effort on bond 2, flow on bond 7, displacement on bond 8 and the power on bond 9. The second plot will go from 24. to 30. in time steps of 1.0 (0.5*2= 1.0), and the ordinate variable is the net energy transferred on bond 9. The printing will list state variables 3 and 4 and source variable 7 at every fifth step.

2.1.7 Display Options

The DISPLAY option allows the user to determine what information will be displayed during execution of the program. By entering the appropriate key-word list following the DISPLAY command, the user can cause selected information to be printed on the line printer or typewriter (teletype). The key-word list must be terminated by a period.

The current status of the DISPLAY options is listed if the key-word 'STATUS' is used under the DISPLAY option. Use of the key-word 'RESET' will reset the DISPLAY options to their default values. Options may be changed as often as desired during a job, in order to achieve the output format required.

System outputs not under DISPLAY control are (1) system error messages, both fatal and non-fatal; (2) automatic page headers; and (3) the echo print of the input file, containing the ENPORT commands. All other system output can be controlled by DISPLAY key-words.

KEY-WORD	MEANING
ANALYSIS	Prints basic graph data, such as number of bonds and number of nodes.
POWERS	Prints power flow direction on each bond.
CAUSALITY	Prints causal assignment on each bond, in order of the node list.
CLASS	Prints results of bond classification, the five categories being internal, independent, dependent, dissipative, and source.
FIELD	Prints the field matrices; i.e., those matrices which describe the input-output relationships of the dependent, independent, and dissipative fields.
JUNCTION	Prints the constraint relations, the system matrix, and the reduced system matrix.
STATE	Prints the A, B, and E matrices (i.e., the state equation).
OUTPUT	Prints the C, D, and F matrices along with the output variable list (i.e., the vector output equation).

RESPONSE	Prints the M and N matrices for the vector difference equation, plus (1) the initial state and input vectors; (2) the simulation start, stop, calculation increment, print increment, and integer multiple time steps; and (3) the state, input, and output variable bounds over the simulated time range.
EIGENVALUES	Prints the eignevalues of the A matrix.
CONTROL	Prints the controllability characteristics of the dependent variables of the system.
OBSERVE	Prints the observability characteristics of the dependent variables of the system.
ECHO	Prints the bond graph, parameter list, source definitions, simulation start, stop, and increment times and state variable initial conditions.
REDUCTION	Prints the junction reduction equations for the state and output equations.
STEADYSTATE	Prints the steady-state values of the state, input, and output variables, if appropriate.
PRINT/PLOT	Transfers printing of print/plot data from the standard output file (line printer or typewriter) to the auxiliary output file (TAPE 5).
OFF	Turns off the listed options. Must appear immediately after DISPLAY.

Default options under DISPLAY are:

ANALYSIS POWERS CLASS STATE OUTPUT STEADYSTATE RESPONSE

Examples of usage.

```
DISPLAY  ECHO  STEADYSTATE .
DISPLAY OFF CONTROL OBSERVE .
```

In addition to the system information printed under DISPLAY control, the user may print comment lines at any point in the output file by beginning each input line with '*' in column 1. The entire line, minus the '*', is printed immediately; then input reading continues with the next line. As many comment lines as are desired may be entered, either singly or in groups. This feature is useful in batch processing only.

Comment lines in the input which are to be totally ignored by the system in both batch and interactive modes may be entered by beginning with '$' in column 1.

2.1.8 ENPORT-4 size limitations

Current size limitations in ENPORT-4 are listed in Appendix D.

2.2 Batch Job Structure

The structure of batch jobs in ENPORT-4 is very flexible, in that recycling among various basic operations is allowed. However, certain restrictions on recycling do exist, due to the logic of problem specification. For example, a solution may not be requested if parameters have not been specified.

The basic command list is

DISPLAY
HEADER
GRAPH
PARAMETERS
SOURCES
VARIABLES
SOLUTION
JOBEND

Relations among these commands are discussed below.

DISPLAY ... may be used to control displayed information whenever ENPORT-4 is expecting a command.

HEADER ... may be used whenever ENPORT-4 is expecting a command. Thus multiple identifiers can be used in sections of the same problem.

GRAPH ... specifies a new line code, and hence a new problem. After GRAPH all other basic operations must be executed as if on a new problem.

PARAMETERS ... assumes that reference is made to the existing bond graph. Only parameter <u>changes</u> need be specified. On the first pass after GRAPH <u>all</u> parameters must be given.

SOURCES ... assumes that reference is made to the existing bond graph. Only source definition <u>changes</u> need be specified. If there are no sources SOURCES, SREND may be omitted.

VARIABLES ... assumes that reference is made to the existing bond graph. Each use of this request establishes a new list of output variables. If no output variables other than state and source variables are required, VARIABLES, VREND need not be used. Each time PARAMETERS is used the VARIABLES list should be respecified.

SOLUTION ... must be used each time a new solution or new output is required. All data required under SOLUTION (time control data, initial conditions, plot and print requests) must be included each time.

JOBEND ... must be used to terminate an ENPORT-4 run. It may be requested at any point (e.g., after PREND, GREND, etc.).

An example of a job deck involving recycling is shown in Figure 2-2. Observe that the order of commands is

DISPLAY, HEADER, GRAPH, PARAMETERS, SOURCES, VARIABLES, SOLUTION,

followed by

HEADER, PARAMETERS (change only), SOURCES (change only), VARIABLES (following use of PARAMETERS), SOLUTION and JOBEND.

```
CONTENTS OF TAPE7

          INFORMATION FOR PROCESSING

CARD  1 = {HEADER
CARD  2 = {PROBLEM UDOT TEST
CARD  3 = {DISPLAY ANALYSIS CLASS STATE OUTPUT .
CARD  4 = {GRAPH
CARD  5 = {SE 5 , 0 5 6 7 , 1 6 1 8 , C 1 , 1 7 4 9 , R 4 , 0 8 9 2 3 , C 2 ,
CARD  6 = {R 3 .
CARD  7 = {GREND
CARD  8 = {PARAMETERS
CARD  9 = {C1= 5.0  R4=   10.0  C2=   15.0  R3=   15.0
CARD 10 = {PREND
CARD 11 = {SOURCES
CARD 12 = {U1 =   CONSTANT  12.0
CARD 13 = {SREND
CARD 14 = {VARIABLES
CARD 15 = {Q1 Q2 PW1 PW2 PW3 PW4 PW5 PW6 PW7 PW8 PW9
CARD 16 = {VREND
CARD 17 = {SOLUTION
CARD 18 = {TIN=  0.0 TFIN=  50.0 DT=  .5
CARD 19 = {ICS=   0.0
CARD 20 = {BPLOT 0.0 0.1 F5 .
CARD 21 = {BPLOT 0.0 10.0 0.1 PW4 Q1 .
CARD 22 = {BPLOT 0.0 10.0 0.1 E1 E2 E3 E4 E5 .
CARD 23 = {SLEND
CARD 24 = {HEADER
CARD 25 = {TEST OF RECURSION
CARD 26 = {PARAMETERS
CARD 27 = {R4= 100.
CARD 28 = {PREND
CARD 29 = {SOURCES
CARD 30 = {E5= SIN   1.0  5.0  0.0
CARD 31 = {SREND
CARD 32 = {VARIABLES
CARD 33 = {PW1
CARD 34 = {VREND
CARD 35 = {SOLUTION
CARD 36 = {TIN= 0.0 TFIN=  1000.   DT=  10.
CARD 37 = {ICS= 0.0
CARD 38 = {BPLOT 0.0 1000. 10.0 E1 F1 .
CARD 39 = {SLEND
CARD 40 = {JOBEND

END-OF-FILE
```

Figure 2-2. Sample deck for batch mode showing recycling.

2.3. ENPORT Batch Error Diagnostic Procedure

As the ENPORT job deck is processed, checks are performed on the validity of the requests. Also the graph structure is analyzed and various system tests are made. If an error is encountered, a message is printed and the program is terminated. In the message an error type number is given. More information about this type of error may be found in Appendix E under this type number. In addition, the routine in which the error occurred or was detected is given. By reference to Appendix F, some debugging help may be obtained. Also with some of the errors, the value of ABORT is of significance and may be helpful in debugging.

EXAMPLE - 1 (see Figure 2-3)

In this job an illegal node name was detected. The "line code being processed" display shows the trouble to be a J 4 in the graph string. ABORT in this case is of no significance.

EXAMPLE - 2 (see Figure 2-4)

In this job there is a flaw in the graph structure. The "line code being processed" is of no help, but the value for ABORT gives the number of the bond which is attached at more than two ends.

NOTE: If blank separators are not properly used, invalid processing and incorrect error diagnostics may be made. For correct spacing, see Section 2.1.

```
CONTENTS OF TAPE 2   (INPUT)
CARD  1 = {HEADER
CARD  2 = {TEST FOR REDUNDENCIES
CARD  3 = {SETUP
CARD  4 = {CAUSAL CONSTIT INFLUENCE REDUCED .
CARD  5 = {GRAPH
CARD  6 = {SE 1 , C 2 , C 3 , J 4 , I 5 , SF 6 , R 7 , SF 8 , R 9 ,
CARD  7 = {0 1 10 12 , 1 2 10 11 , 0 3 11 13 , 1 4 12 14 , 0 14 16 18 ,
CARD  8 = {1 8 16 17 , 0 15 17 18 , 1 5 13 15 , 1 7 18 20 , 1 9 19 26 ,
CARD  9 = {0 6 20 21 ,
CARD 10 = {GREND
CARD 11 = {PARAMETERS
CARD 12 = {C2= 1.0 C3= 1.0 I4= 1.0 I5= 1.0 R7= 1.0 R9= 1.0
CARD 13 = {PREND
CARD 14 = {JOBEND
END-OF-FILE
```

```
SYSTEM BOND GRAPH
NODE    BONDS
 SE        1
 C         2
 C         3
 J         4
```

```
**FATAL ERROR***
TYPE = 103            DETECTED BY RENEL
ILLEGAL NODE NAME, MUST BE  SE,SF,C,I,R,TF,GY,0, OR 1.
CONTENTS OF LINE CODE BEING PROCESSED
SE 1 , C 2 , C 3 , J 4 , I 5 , SF 6 , R 7 , SF 8 , R 9 ,
PROGRAM ABORTED            ABORT =   1
```

Figure 2-3. Example of line code containing illegal node type.

```
CONTENTS OF TAPE2  (INPUT)

CARD  1 = [HEADER
CARD  2 = [TEST FOR REDUNDENCIES
CARD  3 = [SETUP
CARD  4 = [CAUSAL CONSTIT INFLUENCE REDUCED .
CARD  5 = [GRAPH
CARD  6 = [SE 1 , C 2 , C 3 , I 4 , I 5 , SF 6 , R 7 , SF 8 , R 9 ,
CARD  7 = [0 1 10 12 , 1 2 10 11 , 0 3 11 13 , 1 4 12 14 , 0 14 16 18 ,
CARD  8 = [1 8 16 17 , 0 15 17 18 , 1 5 13 15 , 1 7 18 20 , 1 9 19 26 ,
CARD  9 = [0 6 20 21 .
CARD 10 = [GREND
CARD 11 = [PARAMETERS
CARD 12 = [C2= 1.0 C3= 1.0 I4= 1.0 I5= 1.0 R7= 1.0 R9= 1.0
CARD 13 = [PREND
CARD 14 = [JOBEND

END-OF-FILE


SYSTEM BOND GRAPH

NODE     BONDS
 SE         1
 C          2
 C          3
 I          4
 I          5
 SF         6
 R          7
 SF         8
 R          9
 0          1    10    12
 1          2    10    11
 0          3    11    13
 1          4    12    14
 0         14    16    18
 1          8    16    17
 0         15    17    18
 1          5    13    15
 1          7    18    20
 1          9    19    26
 0          6    20    21

GRAPH ANALYSIS.


**FATAL ERROR***

TYPE = 105           DETECTED BY ANAST

BOND CONNECTED TO MORE THAN TWO NODES

CONTENTS OF LINE CODE BEING PROCESSED

0 6 20 21 .

PROGRAM ABORTED            ABORT =  18
```

Figure 2-4. Example of line code containing illegal structure.

2.4. ENPORT Control Cards for the CDC 6500 System

The following list of control cards is required for the ENPORT program operated on the CDC 6500 computer at Michigan State University. These cards are not necessarily valid on other CDC computers. These are the cards in effect on July 20, 1974.

CARD 1: PNC
2: User ID,CM56000,T400,JC500,RG2.
3: PW= password
4: HAL,BANNER, user id.
5: HAL,ENPORT.
6: 7/8/9 multipunch card
.
. (ENPORT job deck)
.
N: 6/7/8/9 multipunch card

Remarks.

1. PNC refers to the problem number card.
2. The user id card is subject to change, and includes
   CM, core memory;
   T, time;
   JC, job cost;
   RG, rate group.
   Other options are available.
3. Changes in the CDC 6500 system operation occasionally require changes in one or more of the control cards. This page should be kept up to date for your local computer installation.

REFERENCES

[1] Karnopp, D. C. and R. C. Rosenberg, System Dynamics: A Unified Approach (Division of Engineering Research, Michigan State University, E. Lansing, Mi., 1971), Chapter 5.

[2] Karnopp, D. C. and R. C. Rosenberg, Analysis and Simulation of Multiport Systems (M. I. T. Press, Cambridge, Mass., 1968), Chapter 5.

[3] Rosenberg, R. C., "State-space formulation for bond graph models of multiport systems," Trans. ASME, J. Dynamic Systems, Measurement, and Control 93, 1, 35-40 (March 1971).

CHAPTER 3. INTERACTIVE ENPORT-4

The interactive version of ENPORT-4 is similar in most respects to the batch version described in Chapter 2. For particular details of usage, including line formats, meaning of commands, etc., refer to Chapter 2.

3.1. Interactive Mode

The use of interactive ENPORT-4 is shown in the series of figures 3-1 through 3-7, which are a complete record of an example. Figure 3-1 shows the log-in procedure for the Michigan State University CDC 6500 computer, and will differ from system to system. Once ENPORT-4 is active it will request a level-1 command. These are

HEADER, GRAPH, PARAMETERS, SOURCES, VARIABLES,
SOLUTION, SETUP, and JOBEND,

and are discussed in Section 2.2.

To enter a bond graph type "GRAPH", as in Figure 3-2. Requests for data made by the program are followed by an asterisk; data entered by the user follows this asterisk. Details of usage for the GRAPH command are given in Section 2.1.2.

Figure 3-3 illustrates the specification and verification of parameters. Figure 3-4 gives some formulation results, and illustrates the interactive use of SOURCES. For details on PARAMETERS, refer to Section 2.1.3; for SOURCES, see Section 2.1.4.

Specifications of output variables and solution control data are shown in Figure 3-5, and follow the usage patterns described in Sections 2.1.5. and 2.1.6, respectively.

At this point, multiple plots and prints are available. Two particular requests are shown in Figures 3-6 and 3-7. Notice that the plotting request is TPLOT (teletype plot) which is sized to fit on the page nicely. Fifty-one abscissa values are used.

To continue with the study, one responds appropriately to the request for a level-1 command. To exit in orderly fashion, one types JOBEND.

```
07/22/74   MSU HUSTLER 2 L239 LSD 34.04  07/17/74

TYPE PASSWØRD, PN, AND USER ID.
XXXXXXXXXX35000,ENPST←YS.

SS73031, LINE  23
LAST ACCESS:  S 07/22/74 10:27
RUNS: 645   BALANCE:  $   91.34
SYSTEM: FØRTRAN LINE RANGE 0 - 0
LENGTH = 72 MARGIN=   LINES=

READY  10:34.35


HAL,ENPØRT.
 HAL 3.11
 EXEC BEGUN.10.34.48


07/22/74 --- M S U   6 5 0 0   E N P Ø R T ---. 10.34.48
```

Figure 3-1. Interactive ENPORT-4 log-in and program retrieval.

```
ENTER LEVEL-1 CØMMAND.
* GRAPH

ENTER SYSTEM BØND GRAPH.

NØDE/BØNDS
* SF 1 , C 2 , 0 1 2 3 , TF 3 4 , 0 4 5 6 , 5 5 ←←←←R 5 , I 6 .

GRAPH ANALYSIS.

 THE GRAPH HAS     7 NØDES AND     6 BØNDS.

ASSIGN PØWER FLØWS.  (YES/NØ)
* NØ

 PØWER FLØWS
BØND  FRØM TØ
  1    SF   0
  2    0    C
  3    0    TF
  4    TF   0
  5    0    R
  6    0    I

ENTER USER CAUSALITY.  (YES/NØ)
* NØ

ASSIGN ANY ACTIVATIØN.  (YES/NØ)
* NØ

 THE NØDE-BØND INCIDENCE MATRIX WITH CAUSALITY
 + = STRØKE END,    - = NØN  STRØKE END.
 1  SF       1   0   0   0   0
 2  C       -2   0   0   0   0
 3  0       -1   2  -3   0   0
 4  TF       3  -4   0   0   0
 5  0        4  -5  -6   0   0
 6  R        5   0   0   0   0
 7  I        6   0   0   0   0
```

Figure 3-2. Interactive graph specification.

```
THERE ARE    4 FIELD BØNDS AND    2 JUNCTION BONDS.

THE BØNDS INTERNAL TØ THE JUNCTIØN STRUCTURE ARE --
  4  3

THE  2 INDEPENDENT ENERGY VARIABLES ARE --

Q 2  P 6

THE  1 DISSIPATIØN VARIABLES ARE --

F 5

THE 1 SØURCE VARIABLES ARE --

F 1


ENTER LEVEL-1 CØMMAND.
* PARAMETERS


ENTER DESIRED PARAMETER VALUES.
* C2= 1.0   TF34= 0.5   R5= 5.0   I6= 0.5    PREND

PARAMETER INITIALIZATIØN/MØDIFICATIØN

C 2
                1.0000E+00
I 6
                5.0000E-01
R 5
                5.0000E+00
TF 3 4
                5.0000E-01
```

Figure 3-3. Interactive parameters specification.

```
THE A MATRIX

 -8.000E-01  -1.000E+00
  2.000E+00   0.

THE B MATRIX

  1.000E+00
  0.

THE EIGENVALUES ØF THE A MATRIX...

NUMBER     REAL PART        IMAGINARY PART   TIME CØNSTANT

   1       -4.0000E-01       1.3565E+00      -2.5000E+00
   2       -4.0000E-01      -1.3565E+00      -2.5000E+00


ENTER LEVEL-1 CØMMAND.
* SØURCES

ENTER SØURCE DEFINITIØN ØR SREND.
* F1=   STEP

ENTER A ØR S  AND TAU.
* 2.5   0.5

ENTER SØURCE DEFINITIØN ØR SREND.
* SREND
```

Figure 3-4. Interactive source specification.

```
ENTER LEVEL-1 CØMMAND.
* VARIABLES

ENTER UP TØ 20 DESIRED ØUTPUTS
* PW3  PW4  VREND

THE  4 ØUTPUT VARIABLES ARE --

F 4   E 4   E 3   F 3

THE C MATRIX.
  0.             5.0000E-01
  2.0000E+00  0.
  1.0000E+00  0.
  0.             1.0000E+00

THE D MATRIX.
  0.
  0.
  0.
  0.


ENTER LEVEL-1 CØMMAND.
* SØLUTIØN

ENTER T-INITIAL, T-FINAL, AND DELTA-T.
* TIN= 0.0   TFIN= 10.0   DT= 0.1

ENTER THE INITIAL CØNDITIØNS FØR THE STATE VARIABLES
* ICS=  0.0   0.0   .

STATE VARIABLE INITIAL CØNDITIØNS
XO( 1) =    0.
XO( 2) =    0.

STARTING AT  0.              101 STAGES IN STEPS ØF
  1.0000E-01 WERE CALCULATED.

SYSTEM VARIABLE BØUNDS

VARIABLE    MAXIMUM      MINIMUM
X( 1) =    1.2079E+00 -4.7868E-01
X( 2) =    3.4897E+00  0.
Y( 1) =    1.7448E+00  0.
Y( 2) =    2.4158E+00 -9.5737E-01
Y( 3) =    1.2079E+00 -4.7868E-01
Y( 4) =    3.4897E+00  0.
U( 1) =    2.5000E+00  0.
```

Figure 3-5. Interactive variables and solution specifications.

```
ENTER TYPE ØF ØUTPUT DESIRED ØR SLEND.
* TPLØT

ENTER T1  T2  INTERVAL FREQUENCY AND VARIABLE LIST.
* 0.0    10.0    .2   Q2   P6  .
         PLØT-*       Q  2 VERSUS TIME.
         PLØT-+       P  6 VERSUS TIME.

        TIME STEP =      2.000E-01
  3.84E+00 +
           I
           I           +++++
           I         ++      ++
           I        +          ++              ++++++++
  2.24E+00 +                     ++      +++++         ++++++++
           I       +               ++++++
           I      +
           I
           I    *+***
  6.40E-01 +   *+     *
           I  *+       **
           +--+------+--*--------+-----**********---+---------*
           +++           **      ****            ************
           I              ******
 -9.60E-01 +
           I
           I
           I
           I
 -2.56E+00 +
           I
           I
           I
           I
 -4.16E+00 +
```

Figure 3-6. Interactive plotting output, example 1.

```
ENTER TYPE ØF ØUTPUT DESIRED ØR SLEND.
* TPLØT

ENTER T1  T2  INTERVAL FREQUENCY AND VARIABLE LIST.
*  0.0   10.0     .1    PW3  F1  .
       PLØT-*      F  1 VERSUS TIME.
       PLØT-+      PW 3 VERSUS TIME.

      TIME STEP =      1.000E-01
```

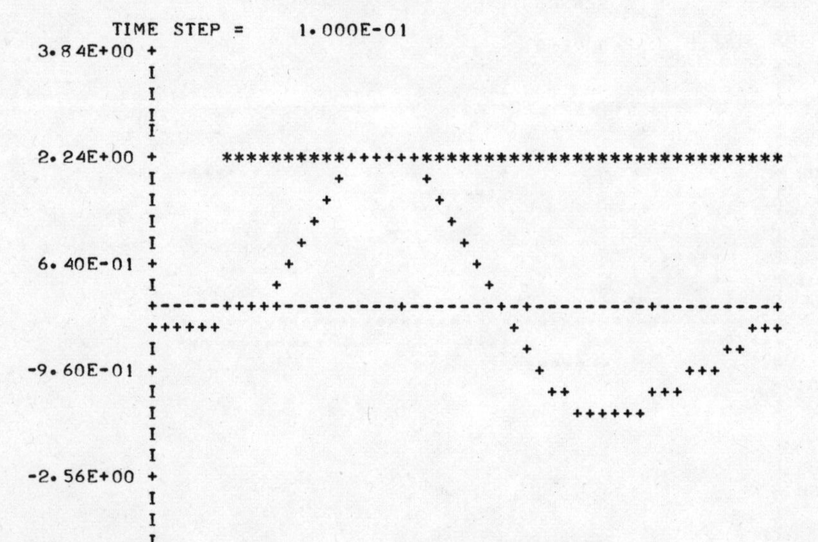


```
ENTER TYPE ØF ØUTPUT DESIRED ØR SLEND.
* SLEND

ENTER LEVEL-1 CØMMAND.
* JØBEND

NØRMAL TERMINATIØN
```

Figure 3-7. Interactive plotting output, example 2.

3.2. ENPORT-4 Interactive Error Procedures

Some errors in ENPORT-4 can be corrected directly while on-line, saving considerable time in obtaining good results. It is impossible to anticipate and illustrate all errors, diagnostics and recovery procedures, but a few are shown in Figures 3-8 and 3-9.

If an error in entering a level-1 command is made, as with 'GRAPH' in Figure 3-8, detection and correction are immediate. If an illegal node is typed in the line code, it is possible to continue from the error or to start the entire code again. An illustration of the continuation procedure is shown in Figure 3-8. Similarly, if an element has too many bonds attached (as for the 1-junction), one may restart or continue.

Errors in entering parameters may be corrected by restarting or continuing as illustrated in Figure 3-9.

Generally, it is advisable to restart the section of processing if these is serious doubt as to when the error occurred, or what data ENPORT now has about the problem.

```
GRAPH

***NØN FATAL ERRØR***

TYPE = 118          DETECTED BY DRIVER

ILLEGAL LEVEL-1 CØMMAND


ENTER LEVEL-1 CØMMAND.
*GRAPH


ENTER SYSTEM BØND GRAPH.


NØDE/BØNDS
*R 1 , Q 3 ,

***NØN FATAL ERRØR***

TYPE = 103          DETECTED BY RENEL

ILLEGAL NØDE NAME, MUST BE  SE,SF,C,I,R,TF,GY,O, ØR 1.


NØDE-BOND SETS PRØCESSED.
 R          1    0    0    0    0

ENTER RESTART ØR CØNTINUE.
*CØNTINUE
*C 3 , 1 1 2 3 4 5 6 , TF 5 8 ,

***NØN FATAL ERRØR***

TYPE = 101          DETECTED BY RGRAPH

MØRE THAN FIVE BØNDS ATTACHED TØ A SINGLE NØDE.


NØDE-BØND SETS PRØCESSED.
 C          3    0    0    0    0
```

Figure 3-8. Interactive recovery from graph errors.

```
ENTER DESIRED PARAMETER VALUES.
*R1= 4.33
*N4=  2.0  TF3←24=  3.RT5  14=  7.0  PREND


***NØN FATAL ERRØR***


TYPE = 142        DETECTED BY RDPAR


ILLEGAL NØDE NAME IN PARAMETER IDENTIFIER.


RE-ENTER PARAMETERS BEGINNING FRØM THE PØINT LISTED ABØVE.
```

```
READY 09.57.24

LØGØUT,T.

JØB CØST:  $        4.96
```

Figure 3-9. Interactive recovery from parameter error and logout.

3.3. Remarks about the MSU CDC 6500 System

For the convenience of the users of the Michigan State University CDC 6500 computer, several relevant features of that interactive system will be summarized here. Some of these features are subject to change for reasons outside our control.

3.3.1. Log-in, execution and log-out

The log-in procedure is illustrated in Figure 3-1. Retrieval and execution is indicated by HAL, ENPORT. HAL refers to the Hustler Auxiliary Library, and indicates that ENPORT is maintained as an auxiliary system library program.

To log-out simply type 'LOGOUT, T.' which drops all unnecessary files and indicates the job cost, as shown in Figure 3-9.

3.3.2. Character, line and program controls

To erase a set of characters in the line being entered, type a set of backspaces (←). For an example, see the 'NODE/BONDS' line in Figure 3-2, which changes '5 5' to 'R 5', including blanks.

To erase a line before it is sent, push (CTRL and X) simultaneously. Several slashes will be printed to indicate deletion of the data.

To abort the program while it is executing, push (ESC). If the program is printing at the time, several pushes may be required. This is a disorderly abort, and one may assume that all data for the problem is lost.

CHAPTER 4. ENPORT-4 PROGRAM DESIGN

This chapter describes the program design for ENPORT-4. It is intended for the general reader of this guide, so that he or she may better appreciate what the program is doing and thereby obtain more use from it. It is also intended as a guide to the program listings, so that a person desiring to study and/or modify the program at the system level will be able to do so.

4. 1. The Basic ENPORT Procedure

The basic ENPORT procedure is shown in Figure 4-1. The first four blocks are the heart of the preliminary processing, which enables the equation formulation to be carried out in a straightforward manner. Required input data is the bond graph model; output is a set of key vectors which are used in subsequent processing.

Blocks (5) through (9) constitute the equation formulation and reduction phase. The several system vectors are related by various matrix equations. These are set up and reduced in a series of operations to obtain the desired state-space form. Required input data are the parameters; available outputs are the state-space matrices A, B and E. A number of intermediate results are available, and are stored for further use in calculating additional results upon request.

Block (10) is used to define the source vector, U, from a list of choices (see Section 2. 1. 4). Block (11) is used to define auxiliary output variables (i. e., ones other than state or source variables) (see Section 2. 1. 5).

Finally, blocks (12-14) represent the solution phase, in which the matrix exponential technique is used (see, for example, reference [1], Chapter 3). This method is not the only available technique for integrating a simultaneous set of linear, constant coefficient, first-order differential equations, but it has several nice features which are important in ENPORT. One is that the method will operate with a singular A-matrix; a second is that the error in the (homogeneous) solution can be bounded in an efficient way. It is of course possible to provide alternative integration methods in blocks (12) and (13), if the user so desires.

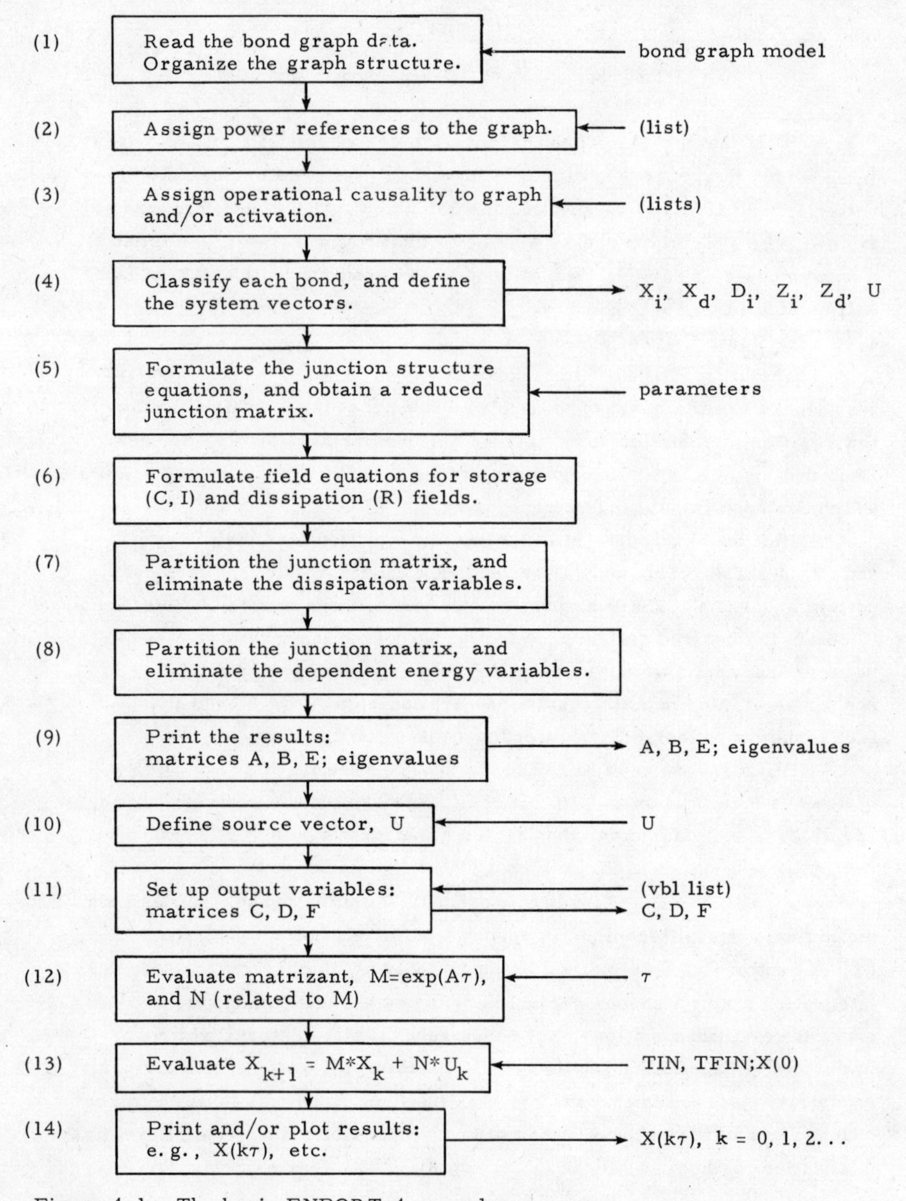

Figure 4-1. The basic ENPORT-4 procedure.

Required data now includes τ, the time step for output, TIN, initial time, and TFIN, the final time, the initial conditions, X(0), and the output requests; output includes the state vector at time intervals of τ, as well as other requested results.

In the following sections, reference will be made to "causality". The following description, although very brief, may help the reader understand the subsequent development. In addition, thorough discussions of causality are to be found in any of the general bond graph references.

Operational causality is a concept of assigning a sense of "cause-effect" or "input-output" to each of the signals (i. e., efforts and flows) in the system. This is done in bond graphs in a unique fashion, in that each bond is marked with a single stroke, showing the causal sense of both signals on that bond simultaneously. The two signals are directed in opposite senses, thus establishing a bilateral action-reaction pattern as shown below. All relations among signals may be organized compatibly before any relations are written by assigning causality to the graph as a whole.

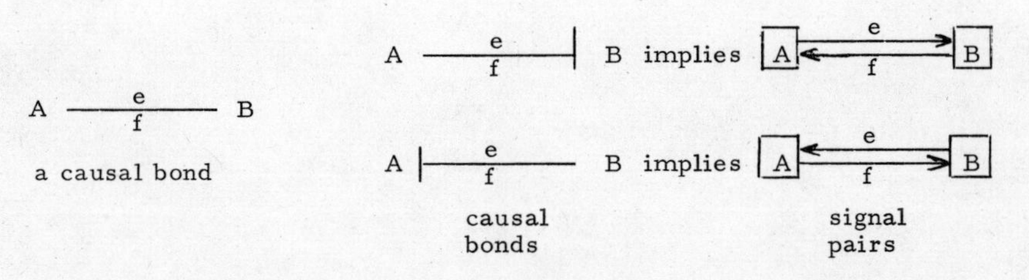

For example, an effort source, SE ——, must always be shown causally as SE ——|, because its effort is imposed by SE on the subsystem to which it is adjoined. Simultaneously the flow is input to the SE from its adjoined subsystem.

A capacitance element, C ——, can have either integration causality, C ——|, or differentiation causality, C |—— . The implied relations for each are

$$C \xrightarrow[f]{e}|$$

$$q = \int f \cdot dt \text{ (input)}$$

(output) $e = (1/C)\, q$

Integration causality

$$C|\xrightarrow[f]{e}$$

$$q = C \cdot e \text{ (input)}$$

(output) $f = dq/dt$

Differentiation causality

4. 2. A Description of the Preliminary Processing Procedure

An extensive discussion of the material in this section can be found in reference [2].

In following a description of the preliminary processing procedure, it will be helpful to keep in mind the picture shown in Figure 4-2. Part (a) is a symbolic bond graph representation of the major fields of a multiport system; namely, the independent and dependent storage fields, the source field, the dissipation field and the junction structure. Notice that each basic multiport type belongs to only one field except for C and I elements. C and I elements are classified according to the causality assigned.

In part (b) of Figure 4-2, two complementary vectors are associated with each field, one vector of inputs and one vector of outputs. The input and output vector elements are ordered corresponding to the bond from which they come. The scalar product of the vectors gives the power associated with the field. For the junction structure, the input vector is the composite of the field output vectors, and the junction structure output vector becomes the input vectors to the various fields, when it is partitioned suitably. Each field and the junction structure is characterized by a matrix, whose size is governed by the number of field ports.

The energy variables, X, are the q's of the C elements and the p's of the I elements. Independent energy variables, X_i, refer to C and I elements with integration causality; dependent energy variables, X_d, refer to C and I elements with differentiation causality. The source vector, U, is composed of the e's on SE elements and f's on SF elements. The principal system vectors are:

- X_i, time derivative of the independent energy variables (X_i);
- Z_i, the independent coenergy variables;
- $\dot{X}_d$, time derivative of the dependent energy variables (X_d);
- Z_d, the dependent coenergy variables;
- U, the source vector;
- V, the source complement vector (of no significance analytically);

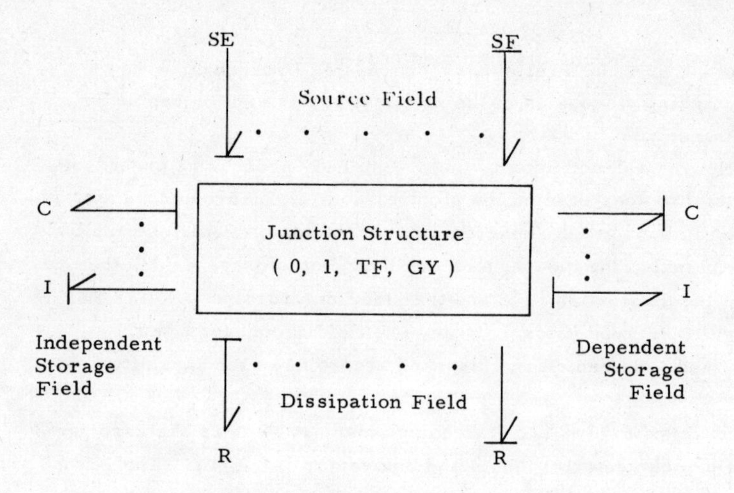

(a) symbolic bond graph representation

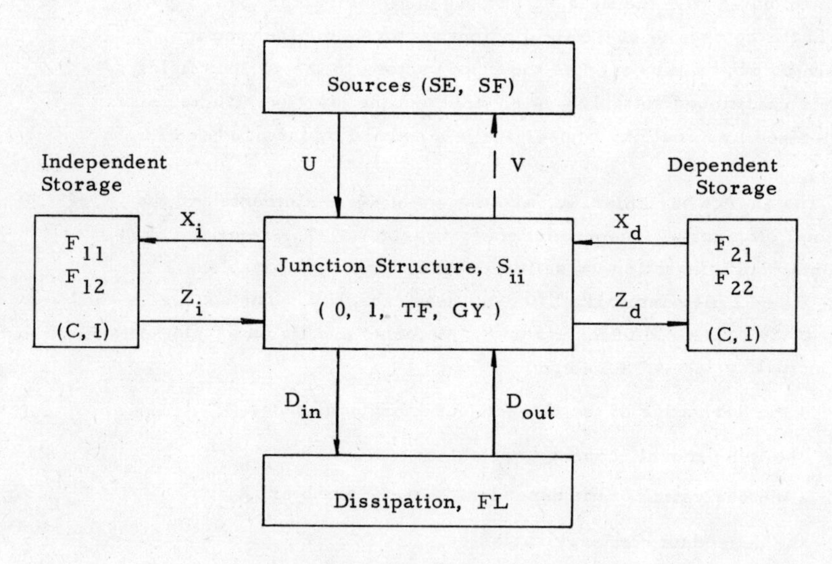

(b) vectors associated with the fields

Figure 4-2. Principal fields in linear multiport systems.

D_i, input vector to the dissipation field; and

D_o, output vector from the dissipation field.

The principal matrices are:

S, the junction structure matrix;

FL, the dissipation matrix; and

F, the storage field matrix (subsequently partitioned into subarrays for the independent and dependent fields).

The initial formulation uses all vectors and all matrices (hence, all implied bond graph variables). The matrix equation is

$$\begin{bmatrix} \dot{X}_i \\ Z_d \\ D_{in} \\ V \end{bmatrix} = \begin{bmatrix} S_{11} & S_{12} & S_{13} & S_{14} \\ S_{21} & S_{22} & S_{23} & S_{24} \\ S_{31} & S_{32} & S_{33} & S_{34} \\ S_{41} & S_{42} & S_{43} & S_{44} \end{bmatrix} \begin{bmatrix} Z_i \\ \dot{X}_d \\ D_{out} \\ U \end{bmatrix} \qquad \text{eq.[4-1]}$$

In addition there are field equations for the storage elements,

$$\begin{bmatrix} Z_i \\ Z_d \end{bmatrix} = \begin{bmatrix} F_{11} & F_{12} \\ F_{21} & F_{22} \end{bmatrix} \begin{bmatrix} X_i \\ X_d \end{bmatrix} \qquad \text{eq.[4-2]}$$

and the dissipation elements,

$$[\, D_{out} \,] = [FL]\, [\, D_{in} \,] \qquad \text{eq.[4-3]}$$

With the aid of causality, the equations are manipulated and variables are eliminated, until the desired state-space form,

$$\dot{X}_i = A\, X_i + B\, U + E\, \dot{U} \,, \qquad \text{eq.[4-4]}$$

is obtained as a result.

Use of the request SETUP (see Section 2.1.7) is recommended for tracing the formulation and reduction steps taken by the program. The print-out is in the form of matrices, implying matrix equations.

For further assistance, Appendix G lists the principal subprograms used in ENPORT-4, classified according to block (see Figure 4-1) with brief comments as appropriate.

4.3. An Example of SETUP Tracing

The basic requests under SETUP for tracing the system procedure and intermediate results are listed in Section 2.1.7. In this section, most of the requests are used in an example. The purpose is to display the type of information keyed to each particular request. References to vectors are made in accordance with the definitions in Figure 4-2.

The job being run is the beam-block transducer system previously considered in Section 1.2.1. In Figure 4-3, cards 3 and 4 contain the SETUP tracing list. Verification of the information to be displayed is given by the DEBUG OPTION STATUS.

Bond graph setup data is shown in Figure 4-4. For convenience the bond graph with powers direction is also shown. The causal matrix at level 1 shows that bond 7 has been causally oriented with the flow into the SE element (6 end) and the effort imposed on the 1 junction (3 end). There are no further consequences of source causality. Causal processing continues with C and I elements. Figure 4-5 shows that causality has been completed at level 2. For example, on bond 1 the flow is directed into the C element and the effort is imposed on the O junction. Results for the entire graph are summarized in the compact standard array next. Then key vectors are identified as shown, where the internal junction structure vector is

$$V_i^t \equiv [E12\ F12\ E11\ F11\ E10\ F10\ E9\ F9\ E8\ F8],$$

the independent energy variable vector is

$$X_i^t \equiv [Q1\ Q2\ P3\ P4\ P5],$$

the output dissipation vector is

$$D_{out} \equiv [E6],$$

and the source vector is

$$U \equiv [E7].$$

```
CONTENTS OF TAPE2  (INPUT)

CARD  1 = {HEADER
CARD  2 = {BEAM BLOCK TRANSDUCER SYSTEM.
CARD  3 = {SETUP CAUSAL CONSTIT INFLUENCE REDUCED FAR FIELD MNMATRICES
CARD  4 = {STEQUATION OPEQUATION .
CARD  5 = {GRAPH
CARD  6 = {C 1 2 , I 3 , I 4 , I 5 , R 6 , 1 7 8 5 6 , GY 8 9 ,
CARD  7 = {1 4 9 10 , 0 10 12 1 , TF 11 12 , 1 2 3 11 , SE 7 .
CARD  8 = {GREND
CARD  9 = {PARAMETERS
CARD 10 = {C12= 7.0 4.0 4.0 9.0  I3=1.0  I4= 2.0  I5= 3.0
CARD 11 = {R6= 6.0  GY89= 5.0  TF1112=  1.5
CARD 12 = {PREND
CARD 13 = {SOURCES
CARD 14 = {E7= CONSTANT  10.0
CARD 15 = {SREND
CARD 16 = {VARIABLES
CARD 17 = {PW3
CARD 18 = {VREND
CARD 19 = {SOLUTION
CARD 20 = {TIN= 0.0  TFIN= 10.0  DT=  0.05
CARD 21 = {ICS= 0.0 0.0 0.0 0.0 0.0
CARD 22 = {TPLOT 0.0 10.0 2 Q1 Q2 P4 .
CARD 23 = {SLEND
CARD 24 = {JOBEND

END-OF-FILE
```

```
SETUP/DEBUG

DEBUG OPTION STATUS

CAUSAL       =  ON
CONSTIT.     =  ON
INFLUENCE    =  ON
REDUCED      =  ON
FAR          =  ON
FIELD        =  ON
MNMATRICES   =  ON
STEQUATION   =  ON
PAR          =  OFF
OPEQUATION   =  ON
CONTROL      =  OFF
OBSERVE      =  OFF
```

Figure 4-3. Beam-block transducer example, showing input and setup status.

SYSTEM BOND GRAPH

NODE	BONDS			
C	1	2		
I	3			
I	4			
I	5			
R	6			
1	7	8	5	6
GY	8	9		
1	4	9	10	
0	10	12	1	
TF	11	12		
1	2	3	11	
SE	7			

GRAPH ANALYSIS.

THE GRAPH HAS 12 NODES AND 12 BONDS.

POWER FLOWS

BOND	FROM	TO
1	0	→ C
2	1	→ C
3	1	→ I
4	1	→ I
5	1	→ I
6	1	→ R
7	SE	→ 1
8	1	→ GY
9	GY	→ 1
10	1	→ 0
11	TF	→ 1
12	0	→ TF

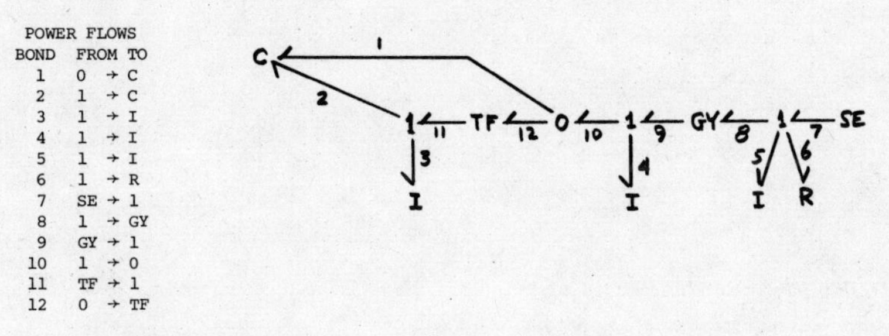

THE CAUSAL MATRIX AT LEVEL 1

		1	2	3	4	5	6	7	8	9	10	11	12
C	-	1	1	0	0	0	0	0	0	0	0	0	0
I	-	0	0	1	0	0	0	0	0	0	0	0	0
I	-	0	0	0	1	0	0	0	0	0	0	0	0
I	-	0	0	0	0	1	0	0	0	0	0	0	0
R	-	0	0	0	0	0	1	0	0	0	0	0	0
1	-	0	0	0	0	1	1	3	1	0	0	0	0
GY	-	0	0	0	0	0	0	0	1	1	0	0	0
1	-	0	0	0	1	0	0	0	0	1	1	0	0
0	-	1	0	0	0	0	0	0	0	0	1	0	1
TF	-	0	0	0	0	0	0	0	0	0	0	1	1
1	-	0	1	1	0	0	0	0	0	0	0	1	0
SE	-	0	0	0	0	0	0	6	0	0	0	0	0

Figure 4-4. Bond graph structure, powers and causality through level 1.

```
M S U   6 5 0 0   ENPORT             VERSION 4.1            UPDATE 50772
------------------------------------------------------------------------------------------------


THE CAUSAL MATRIX AT LEVEL   2

          1  2  3  4  5  6  7  8  9  10  11  12
       +--------------------------------------
  C    -  6  6  0  0  0  0  0  0  0   0   0   0
  I    -  0  0  3  0  0  0  0  0  0   0   0   0
  I    -  0  0  0  3  0  0  0  0  0   0   0   0
  I    -  0  0  0  0  3  0  0  0  0   0   0   0
  R    -  0  0  0  0  0  6  0  0  0   0   0   0
  1    -  0  0  0  0  6  3  3  3  0   0   0   0
  GY   -  0  0  0  0  0  0  0  6  6   0   0   0
  1    -  0  0  0  6  0  0  0  0  3   3   0   0
  0    -  3  0  0  0  0  0  0  0  0   6   0   6
  TF   -  0  0  0  0  0  0  0  0  0   0   6   3
  1    -  0  3  6  0  0  0  0  0  0   0   3   0
  SE   -  0  0  0  0  0  0  6  0  0   0   0   0

THE NODE-BOND INCIDENCE MATRIX WITH CAUSALITY
+ = STROKE END,    - = NON STROKE END.
 1 C       -1  -2   0   0   0
 2 I        3   0   0   0   0
 3 I        4   0   0   0   0
 4 I        5   0   0   0   0
 5 R       -6   0   0   0   0
 6 1        7   8  -5   6   0
 7 GY      -8  -9   0   0   0
 8 1       -4   9  10   0   0
 9 0      -10 -12   1   0   0
10 TF     -11  12   0   0   0
11 1        2  -3  11   0   0
12 SE      -7   0   0   0   0

THERE ARE   7 FIELD BONDS AND   5 JUNCTION BONDS.

THE BONDS INTERNAL TO THE JUNCTION STRUCTURE ARE --
 12  11  10  9  8

THE  5 INDEPENDENT ENERGY VARIABLES ARE --

Q 1  Q 2  P 3  P 4  P 5

THE  1 DISSIPATION VARIABLES ARE --

E 6

THE 1 SOURCE VARIABLES ARE --

E 7
```

Figure 4-5. Causality through level 2 and bond classification.

By implication, the independent coenergy vector is

$$Z_i^t \equiv [E1\ E2\ F3\ F4\ F5] ,$$

the input dissipation vector is

$$D_{in} \equiv [F6],$$

the dependent energy vector is

$$X_d \equiv [0] ,$$

the dependent coenergy vector is

$$Z_d \equiv [0],$$

and the source input vector is

$$V \equiv [F7].$$

The first part of Figure 4-6 shows the parameters as they were interpreted in standard form. Next, each junction structure element has its equations organized according to its power and causal orientations. For example, the first 1-junction (1 7 8 5 6) has the implied matrix equation

$$\begin{bmatrix} F7 \\ F8 \\ E5 \\ F6 \end{bmatrix} = \begin{bmatrix} 0. & 0. & 1. & 0. \\ 0. & 0. & 1. & 0. \\ 1. & -1. & 0. & -1. \\ 0. & 0. & 1. & 0 \end{bmatrix} \begin{bmatrix} E7 \\ E8 \\ F5 \\ E6 \end{bmatrix} .$$

For convenience, the augmented junction is displayed next to the output data. Similarly, node GY 8 9 has been set up with the matrix equation

```
M S U   6 5 0 0   ENPORT            VERSION 4.1          UPDATE  50772
-----------------------------------------------------------------------------

PARAMETER INITIALIZATION/MODIFICATION

C 1 2
                    7.0000E+00  4.0000E+00
                    4.0000E+00  9.0000E+00

I 3
                    1.0000E+00

I 4
                    2.0000E+00

I 5
                    3.0000E+00

R 6
                    6.0000E+00

GY 8 9
                    5.0000E+00

TF1112
                    1.5000E+00

*******************************************************

NODE 1    HAS    4 PORTS

MATRIX INDICES ARE (- IF FLOW OUTPUT)...
      -7  -8   5  -6

   0.          0.           .100E+01   0.

   0.          0.           .100E+01   0.

   .100E+01   -.100E+01    0.         -.100E+01

   0.          0.           .100E+01   0.

*******************************************************

NODE GY   HAS    2 PORTS

MATRIX INDICES ARE (- IF FLOW OUTPUT)...
       8   9

   0.          .500E+01

   .500E+01   0.

*******************************************************
```


Figure 4-6. Parameters and start of junction element setup.

$$\begin{bmatrix} E8 \\ E9 \end{bmatrix} \begin{bmatrix} 0. & 5. \\ 5. & 0. \end{bmatrix} \begin{bmatrix} F8 \\ F9 \end{bmatrix},$$

and the augmented element is shown on the figure. The remaining junction structure elements have their equations implied in Figure 4-7.

Figure 4-8 shows the matrices for the junction structure equation

$$\begin{bmatrix} V_{out} \\ V_i \end{bmatrix} = \begin{bmatrix} S1 & S2 \\ S3 & S4 \end{bmatrix} \begin{bmatrix} V_{in} \\ V_i \end{bmatrix},$$

where V_{out} is the output vector from the junction structure,

V_{in} is the input vector to the junction structure and

V_i is the internal vector.

In this example, these vectors are

$$V_{out} = \begin{bmatrix} F1 \\ F2 \\ E3 \\ E4 \\ E5 \\ F6 \\ F7 \end{bmatrix}, \quad V_{in} = \begin{bmatrix} E1 \\ E2 \\ F3 \\ F4 \\ F5 \\ E6 \\ E7 \end{bmatrix},$$ and V_i has been defined previously.

Next, the internal vector is expressed in terms of the input vector, as shown in Figure 4-9 in the influence matrix. For example, E9 = 5. *F5, and E10 = E1. Then the output vector is expressed in terms of the input vector by the reduced junction structure array. For example, F7 = F5. The next set of data shows the assignment of storage field (C and I) parameters to the field matrix.

In Figure 4-10, the storage field matrix, F11, is given, implying

$$Z_i = [F11]\, X_i ,$$

and the dissipation field matrix is given, implying

```
M S U   6 5 0 0   ENPORT            VERSION 4.1          UPDATE 50772
------------------------------------------------------------------------------------------


NODE 1   HAS   3 PORTS

MATRIX INDICES ARE (- IF FLOW OUTPUT)...
      4  -9  -10

   0.           .100E+01    -.100E+01

    .100E+01   0.           0.

    .100E+01   0.           0.
*************************************************

NODE 0   HAS   3 PORTS

MATRIX INDICES ARE (- IF FLOW OUTPUT)...
     10  12  -1

   0.           0.           .100E+01

   0.           0.           .100E+01

    .100E+01   -.100E+01    0.
*************************************************

NODE TF  HAS   2 PORTS

MATRIX INDICES ARE (- IF FLOW OUTPUT)...
     11 -12

   0.           .150E+01

    .150E+01   0.
*************************************************

NODE 1   HAS   3 PORTS

MATRIX INDICES ARE (- IF FLOW OUTPUT)...
     -2   3  -11

   0.           .100E+01    0.

   -.100E+01   0.           .100E+01

   0.           .100E+01    0.
*************************************************
```

Figure 4-7. Completion of junction element setup.

```
M S U   6 5 0 0   ENPORT              VERSION 4.1            UPDATE  50772
---------------------------------------------------------------------------


S1 MATRIX.
 0.00 0.00 0.00 0.00 0.00 0.00 0.00
 0.00 0.00 1.00 0.00 0.00 0.00 0.00
 0.00-1.00 0.00 0.00 0.00 0.00 0.00
 0.00 0.00 0.00 0.00 0.00 0.00 0.00
 0.00 0.00 0.00 0.00 0.00-1.00 1.00
 0.00 0.00 0.00 0.00 1.00 0.00 0.00
 0.00 0.00 0.00 0.00 1.00 0.00 0.00

S2 MATRIX.
 0.00-1.00 0.00 0.00 0.00 1.00 0.00 0.00 0.00 0.00
 0.00 0.00 0.00 0.00 0.00 0.00 0.00 0.00 0.00 0.00
 0.00 0.00 1.00 0.00 0.00 0.00 0.00 0.00 0.00 0.00
 0.00 0.00 0.00 0.00-1.00 0.00 1.00 0.00 0.00 0.00
 0.00 0.00 0.00 0.00 0.00 0.00 0.00 0.00-1.00 0.00
 0.00 0.00 0.00 0.00 0.00 0.00 0.00 0.00 0.00 0.00
 0.00 0.00 0.00 0.00 0.00 0.00 0.00 0.00 0.00 0.00

S3 MATRIX.
 1.00 0.00 0.00 0.00 0.00 0.00 0.00
 0.00 0.00 0.00 0.00 0.00 0.00 0.00
 0.00 0.00 0.00 0.00 0.00 0.00 0.00
 0.00 0.00 1.00 0.00 0.00 0.00 0.00
 1.00 0.00 0.00 0.00 0.00 0.00 0.00
 0.00 0.00 0.00 1.00 0.00 0.00 0.00
 0.00 0.00 0.00 0.00 0.00 0.00 0.00
 0.00 0.00 0.00 1.00 0.00 0.00 0.00
 0.00 0.00 0.00 0.00 0.00 0.00 0.00
 0.00 0.00 0.00 0.00 1.00 0.00 0.00

S4 MATRIX.
 0.00 0.00 0.00 0.00 0.00 0.00 0.00 0.00 0.00 0.00
 0.00 0.00 0.00 1.50 0.00 0.00 0.00 0.00 0.00 0.00
 1.50 0.00 0.00 0.00 0.00 0.00 0.00 0.00 0.00 0.00
 0.00 0.00 0.00 0.00 0.00 0.00 0.00 0.00 0.00 0.00
 0.00 0.00 0.00 0.00 0.00 0.00 0.00 0.00 0.00 0.00
 0.00 0.00 0.00 0.00 0.00 0.00 0.00 0.00 0.00 0.00
 0.00 0.00 0.00 0.00 0.00 0.00 0.00 0.00 0.00 5.00
 0.00 0.00 0.00 0.00 0.00 0.00 0.00 0.00 0.00 0.00
 0.00 0.00 0.00 0.00 0.00 0.00 0.00 5.00 0.00 0.00
 0.00 0.00 0.00 0.00 0.00 0.00 0.00 0.00 0.00 0.00
```

Figure 4-8. Junction structure equation setup.

```
*************************************************
THE INFLUENCE MATRIX.

( COLS=INPUTS TO JUNCTION STRUCTURE )

      E 1. E 2. F 3. F 4. F 5. E 6. E 7.
     +-----------------------------------
E12.= 1.00 0.00 0.00 0.00 0.00 0.00 0.00
F12.= 0.00 0.00 1.50 0.00 0.00 0.00 0.00
E11.= 1.50 0.00 0.00 0.00 0.00 0.00 0.00
F11.= 0.00 0.00 1.00 0.00 0.00 0.00 0.00
E10.= 1.00 0.00 0.00 0.00 0.00 0.00 0.00
F10.= 0.00 0.00 0.00 1.00 0.00 0.00 0.00
E 9.= 0.00 0.00 0.00 0.00 5.00 0.00 0.00
F 9.= 0.00 0.00 0.00 1.00 0.00 0.00 0.00
E 8.= 0.00 0.00 0.00 5.00 0.00 0.00 0.00
F 8.= 0.00 0.00 0.00 0.00 1.00 0.00 0.00

THE REDUCED JUNCTION MATRIX.

( COLS=INPUTS TO JUNCTION STRUCTURE )

(ROWS=OUTPUTS FROM JUNCTION STRUCTURE )

      E 1. E 2. F 3. F 4. F 5. E 6. E 7.
     +-----------------------------------
F 1.= 0.00 0.00-1.50 1.00 0.00 0.00 0.00
F 2.= 0.00 0.00 1.00 0.00 0.00 0.00 0.00
E 3.= 1.50-1.00 0.00 0.00 0.00 0.00 0.00
E 4.= 1.00 0.00 0.00 0.00 5.00 0.00 0.00
E 5.= 0.00 0.00 0.00-5.00 0.00 1.00 1.00
F 6.= 0.00 0.00 0.00 0.00 1.00 0.00 0.00
F 7.= 0.00 0.00 0.00 0.00 1.00 0.00 0.00

*************************************************

ROW/COL INDICES FOR NODE     1 ARE    1   2
   7.000E+00   4.000E+00
   4.000E+00   9.000E+00

ROW/COL INDICES FOR NODE     2 ARE    3
   1.000E+00

ROW/COL INDICES FOR NODE     3 ARE    4
   2.000E+00

ROW/COL INDICES FOR NODE     4 ARE    5
   3.000E+00

ROW/COL INDICES FOR NODE     5 ARE    6
   6.000E+00
```

Figure 4-9. The influence matrix, reduced junction structure matrix, and storage element assignments.

$$D_{out} = [FL]\ D_{in}\ .$$

Next come a series of reduction steps in which specific vectors are eliminated to arrive at the state-space form. Figures 4-10 and 4-11 show the matrix equations and appropriate arrays. The final result is, as expected, A and B matrices relating X_i and U.

When output variables are requested by VARIABLES (see Section 2. 1. 5), information about their definitions and setup are given as in Figure 4-12, culminating in the C and D matrices. Note that 'X' stands for X_i in the Z2 equation (except where it is used to denote multiplication). Finally, the M and N matrices are given upon request, as shown in Figure 4-12.

It should be clear from this example that very close inspection of the setup procedure is possible by selective use of the SETUP list.

```
F11 MATRIX
  7.0000E+00  4.0000E+00  0.          0.          0.
  4.0000E+00  9.0000E+00  0.          0.          0.
  0.          0.          1.0000E+00  0.          0.
  0.          0.          0.          2.0000E+00  0.
  0.          0.          0.          0.          3.0000E+00

FL MATRIX
  6.0000E+00
```

```
DI = S31*ZI + S32*D(XD)/DT + S34*U

S31 ARRAY
 0.         0.          0.          0.         .100E+01

S32 ARRAY
 0.

S34 ARRAY
 0.
```

```
D(XI)/DT = S11*ZI + S12*D(XD)/DT + S14*U

S11 ARRAY
 0.         0.          -.150E+01   .100E+01  0.
 0.         0.           .100E+01  0.         0.
  .150E+01 -.100E+01    0.         0.         0.
 -.100E+01  0.          0.         0.          .500E+01
 0.         0.          0.         -.500E+01 -.600E+01

S12 ARRAY
 0.
 0.
 0.
 0.
 0.

S14 ARRAY
 0.
 0.
 0.
 0.
  .100E+01
```

```
XD = F21*XI + S21*ZI + S24*U

F21 ARRAY
 0.         0.          0.          0.         0.

S21 ARRAY
 0.         0.          0.          0.         0.

S24 ARRAY
 0.
```

Figure 4-10. Storage and dissipation field matrices, and partially formulated results.

```
D(XI)/DT = S11*XI + S14*U +S12*D(XD)/DT

S11 ARRAY
 0.        0.        -.150E+01  .200E+01 0.
 0.        0.         .100E+01 0.        0.
  .650E+01 -.300E+01 0.        0.        0.
 -.700E+01 -.400E+01 0.        0.         .150E+02
 0.        0.        0.        -.100E+02 -.180E+02

S14 ARRAY
 0.
 0.
 0
 0.
  .100E+01

S12 ARRAY
 0.
 0.
 0.
 0.
 0.
```

```
THE A MATRIX

  0.          0.          -1.500E+00   2.000E+00   0.
  0.          0.           1.000E+00   0.          0.
  6.500E+00  -3.000E+00    0.          0.          0.
 -7.000E+00  -4.000E+00    0.          0.          1.500E+01
  0.          0.           0.         -1.000E+01  -1.800E+01

THE B MATRIX
 0.
 0.
 0.
 0.
 1.000E+00

 THE EIGENVALUES OF THE A MATRIX...

 NUMBER    REAL PART        IMAGINARY PART   TIME CONSTANT
     1     -8.0581E+00       8.2400E+00       -1.2410E-01
     2     -8.0581E+00      -8.2400E+00       -1.2410E-01
     3     -4.3948E-01       3.5333E+00       -2.2754E+00
     4     -4.3948E-01      -3.5333E+00       -2.2754E+00
     5     -1.0048E+00       0.               -9.9521E-01
```

Figure 4-11. Completion of formulation results.

```
M S U   6 5 0 0   ENPORT            VERSION 4.1           UPDATE  50772
--------------------------------------------------------------------------------

SOURCE TABLE.

U 1 = E 7      CONSTANT  VALUE =     .1000E+02

THE  2 OUTPUT VARIABLES ARE --

E 3   F 3

Z2 = ZIC X XI + ZID X U

THE ZIC MATRIX.
  7.0000E+00  4.0000E+00  0.          0.          0.
  4.0000E+00  9.0000E+00  0.          0.          0.
  0.          0.          1.0000E+00  0.          0.
  0.          0.          0.          2.0000E+00  0.
  0.          0.          0.          0.          3.0000E+00

THE ZID MATRIX.
  0.
  0.
  0.
  0.
  0.

THE C MATRIX.
  6.5000E+00 -3.0000E+00  0.          0.          0.
  0.          0.          1.0000E+00  0.          0.

THE D MATRIX.
  0.
  0.

THE M MATRIX...
    .9709E+00   -.4091E-02   -.7434E-01    .9406E-01    .2741E-01
    .8080E-02    .9962E+00    .4974E-01    .2656E-03    .4226E-04
    .3214E+00   -.1503E+00    .9841E+00    .1574E-01    .3223E-02
   -.3297E+00   -.1893E+00    .7869E-02    .8457E+00    .4612E+00
    .6402E-01    .3668E-01   -.1074E-02   -.3075E+00    .3050E+00

THE N MATRIX...
    .4960E-03   0.
    .4361E-06   0.
    .4226E-04   0.
    .1374E-01   0.
    .3098E-01   0.
```

Figure 4-12. Output matrices and the M and N matrices.

REFERENCES

[1] Tou, J. T., Modern Control Theory (McGraw-Hill, New York, 1964).

[2] Rosenberg, R. C., "State-space formulation for bond graph models of multiport systems," Trans. ASME, J. Dynamic Systems, Measurement, and Control, 35-40 (March 1971).

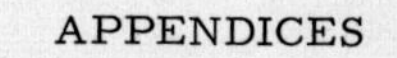
APPENDICES

APPENDIX A

A DEFINITION OF THE BOND GRAPH LANGUAGE

Ronald C. Rosenberg
Associate Professor
Department of Mechanical Engineering
Michigan State University
East Lansing, Michigan 48823

Dean C. Karnopp
Professor
Department of Mechanical Engineering
University of California
Davis, California 95616

1. INTRODUCTION

The purpose of this paper is to present the basic definitions of the bond graph language in a compact but general form. The language presented herein is a formal mathematical system of definitions and symbolism. The descriptive names are stated in terms related to energy and power, because that is the historical basis of the multiport concept.

It is important that the fundamental definitions of the language be standardized because an increasing number of people around the world are using and developing the bond graph language as a modeling tool in relation to multiport systems. A common set of reference definitions will be an aid to all in promoting ease of communication.

Some care has been taken from the start to construct definitions and notation which are helpful in communicating with digital computers through special programs, such as ENPORT, [5]. It is hoped that any subsequent modifications and extensions to the language will give due consideration to this goal.

Principal sources of extended descriptions of the language and physical applications and interpretations will be found in Paynter, [1], Karnopp and Rosenberg, [2, 3], and Takahashi, et al., [4]. This paper is the most highly codified version of language definition, drawing as it does upon all previous efforts.

2. BASIC DEFINITIONS

3. 1. Multiport elements, ports, and bonds

Multiport elements are the nodes of the graph, and are designated by alpha-numeric characters. They are referred to as elements, for convenience. For example, in fig. 1(a) two multiport elements, 1 and R, are shown. Ports of a multiport element are designated by line segments incident on the element at one end. Ports are places where the element can interact with its environment.

For example, in fig. 1(b) the 1 element has three ports and the R element has one port. We say that the 1 element is a 3-port, and the R element is a 1-port.

Bonds are formed when pairs of ports are joined. Thus bonds are connections between pairs of multiport elements.

For example, in fig. 1(c) two ports have been joined, forming a bond between the 1 and the R.

2.2. Bond graphs

A bond graph is a collection of multiport elements bonded together. In the general sense it is a linear graph whose nodes are multiport elements and whose branches are bonds.

A bond graph may have one part or several parts, may have no loops or several loops, and in general has the characteristics of any linear graph.

An example of a bond graph is given in fig. 2. In part (a) a bond graph with seven elements and six bonds is shown. In part (b) the same graph has had its powers directed and bonds labeled.

A bond graph fragment is a bond graph not all of whose ports have been paired as bonds.

An example of a bond graph fragment is given in fig. 1(c), which has one bond and two open, or unconnected, ports.

2.3. Port variables

Associated with a given port are three direct and three integral quantities.

Effort, e(t), and flow, f(t), are directly associated with a given port, and are called the port power variables. They are assumed to be scalar functions of an independent variable (t).

Power, P(t), is found directly from the scalar product of effort and flow, as

$$P(t) = e(t) \cdot f(t).$$

The direction of positive power is indicated by a half-arrow on the bond.

Momentum, $p(t)$, and displacement, $q(t)$, are related to the effort and flow at a port by integral relations. That is,

$$p(t) = p(t_o) + \int_{t_o}^{t} e(\lambda)\, d\lambda$$

and $q(t) = q(t_o) + \int_{t_o}^{t} f(\lambda)\, d\lambda$, respectively.

Momentum and displacement are sometimes referred to as energy variables.

Energy, $E(t)$, is related to the power at a port by

$$E(t) = E(t_o) + \int_{t_o}^{t} P(\lambda)\, d\lambda .$$

The quantity $E(t) - E(t_o)$ represents the net energy transferred through the port in the direction of the half-arrow (i. e., positive power) over the interval (t_o, t).

In common bond graph usage the effort and the flow are often shown explicitly next to the port (or bond). The power, displacement, momentum, and energy quantities are all implied.

2. 4. Basic multiport elements

There are nine basic multiport elements, grouped into four categories according to their energy characteristics. These elements and their definitions are summarized in fig. 3.

Sources.

Source of effort, written SE $\overset{e}{\rightharpoonup}$, is defined by $e = e(t)$.

Source of flow, written SF $\underset{f}{\rightharpoonup}$, is defined by $f = f(t)$.

Storages.

Capacitance, written $\overset{e}{\underset{f}{\rightharpoonup}}$ C, is defined by

$$e = \Phi(q) \text{ and } q(t) = q(t_o) + \int_{t_o}^{t} f(\lambda)\, d\lambda .$$

That is, the effort is a static function of the displacement and the displacement is the time integral of the flow.

Inertance, written $\frac{e}{f}$ I, is defined by

$$f = \Phi(p) \text{ and } p(t) = p(t_o) + \int_{t_o}^{t} e(\lambda)d\lambda .$$

That is, the flow is a static function of the momentum and the momentum is the time integral of the effort.

Dissipation.

Resistance, written $\frac{e}{f}$ R, is defined by

$$\Phi(e, f) = 0.$$

That is, a static relation exists between the effort and flow at the port.

Junctions: 2-port.

Transformer, written $\frac{e_1}{f_1}$ TF $\frac{e_2}{f_2}$, is a linear 2-port element defined by

$$e_1 = m \cdot e_2$$

and $$m \cdot f_1 = f_2,$$

where m is the modulus.

Gyrator, written $\frac{e_1}{f_1}$ GY $\frac{e_2}{f_2}$, is a linear 2-port element defined by

$$e_1 = r \cdot f_2$$

and $$e_2 = r \cdot f_1,$$

where r is the modulus.

Both the transformer and gyrator preserve power (i. e., $P_1 = P_2$ in each case shown), and they must each have two ports, so they are called essential 2-port junctions.

Junctions: 3-port.

Common effort junction, written $\xrightarrow{1} 0 \xleftarrow{3}$ (with bond 2 ↑ below 0)

is a linear 3-port element defined by

$$e_1 = e_2 = e_3 \qquad \text{(common effort)}$$

and $$f_1 + f_2 - f_3 = 0. \qquad \text{(flow summation)}$$

Other names for this element are the flow junction and the zero junction.

Common flow junction, written $\xrightarrow{1} 1 \xrightarrow{3}$, with bond 2 ↑ entering the junction from below,

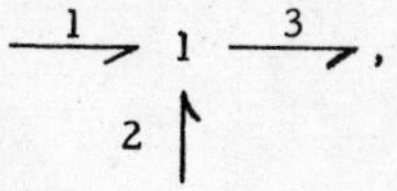

is a linear 3-port element defined by

$$f_1 = f_2 = f_3 \qquad \text{(common flow)}$$

and $$e_1 + e_2 - e_3 = 0. \qquad \text{(effort summation)}$$

Other names for this element are the effort junction and the one junction.

Both the common effort junction and the common flow junction preserve power (i. e., the net power in is zero at all times), so they are called junctions. If the reference power directions are changed the signs on the summation relation must change accordingly.

3. EXTENDED DEFINITIONS

3. 1. Multiport fields

3. 1. 1. Storage fields

Multiport capacitances, or C-fields, are written $\xrightarrow{1} C \xleftarrow{n}$, with bonds 2 ↑ ⋰ entering from below, and characterized by

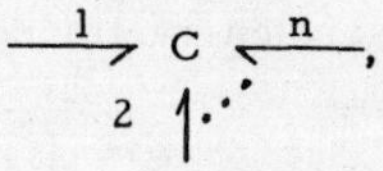

$$e_i = \Phi_i(q_1, q_2, \ldots q_n), \; i = 1 \text{ to } n,$$

and $$q_i(t) = q_i(t_o) + \int_{t_o}^{t} f_i(\lambda)\, d\lambda, \; i = 1 \text{ to } n.$$

Multiport inertances, or I-fields, are written $\xrightarrow{1} I \xleftarrow{n}$, with bonds 2 ↑ ⋰ entering from below, and characterized by

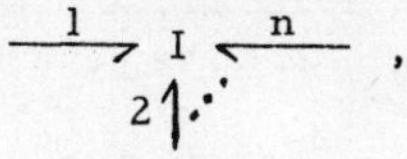

$$f_i = \Phi_i(p_1, p_2, \ldots p_n), \; i = 1 \text{ to } n,$$

and $$p_i(t) = p_i(t_o) + \int_{t_o}^{t} e_i(\lambda)\, d\lambda, \; i = 1 \text{ to } n.$$

If a C-field or I-field is to have an associated 'energy' state function then certain integrability conditions must be met by the Φ_i

functions. In multiport terms the relations given above are sufficient to define a C-field and I-field, respectively.

Mixed multiport storage fields can arise when both C- and I-type storage effects are present simultaneously. The symbol for such an element consists of a set of C's and I's with appropriate ports indicated.

For example, $\xrightarrow{1} \text{ICI} \xleftarrow{3}$ (with port 2 arrow pointing up into ICI) indicates the existence of a set of relations

$$f_1 = \Phi_1 (p_1, q_2, p_3),$$

$$e_2 = \Phi_2 (p_1 q_2, p_3),$$

$$f_3 = \Phi_3 (p_1, q_2, p_3),$$

and

$$p_1(t) = p_1(t_o) + \int_{t_o}^{t} e_1(\lambda)\, d\lambda ,$$

$$q_2(t) = q_2(t_o) + \int_{t_o}^{t} f_2(\lambda)\, d\lambda ,$$

$$p_3(t) = p_3(t_o) + \int_{t_o}^{t} e_3(\lambda)\, d\lambda .$$

Multiport dissipators, or R-fields, are written $\xrightarrow{1} \text{R} \xleftarrow{n}$ (with port 2 arrow pointing up into R, ...) and are characterized by

$$\Phi_i(e_1, f_1, e_2, f_2, \ldots e_n, f_n) = 0, \; i = 1 \text{ to } n.$$

If the R-field is to represent pure dissipation, then the power function associated with the R-field must be positive definite.

Multiport junctions include 0 junctions and 1 - junctions with n ports, n≥2. The general case for each junction is given below.

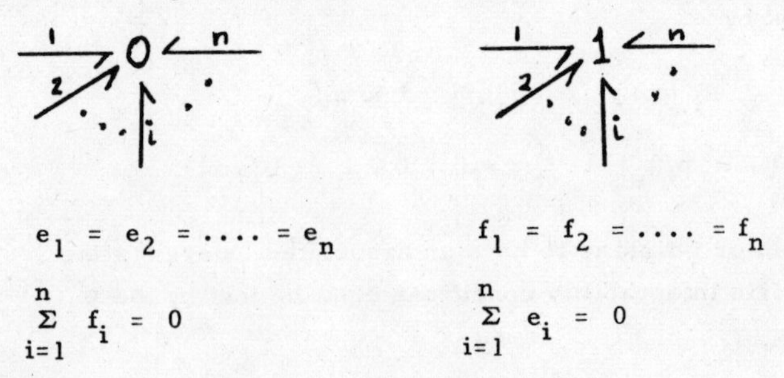

$$e_1 = e_2 = \ldots = e_n \qquad\qquad f_1 = f_2 = \ldots = f_n$$

$$\sum_{i=1}^{n} f_i = 0 \qquad\qquad \sum_{i=1}^{n} e_i = 0$$

3.2. Modulated 2-port junctions.

The modulated transformer, or MTF, written $\xrightarrow{1} \overset{\downarrow m(\underline{x})}{\text{MTF}} \xrightarrow{2}$ implies the relations

$$e_1 = m(\underline{x}) \cdot e_2$$

and $\quad m(\underline{x}) \cdot f_1 = f_2,$

where $m(\underline{x})$ is a function of a set of variables, $\underline{x}$. The modulated transformer preserves power; i.e., $P_1(t) = P_2(t)$.

The modulated gyrator, or MGY, written $\xrightarrow{1} \overset{\downarrow r(\underline{x})}{\text{MGY}} \xrightarrow{2}$ implies the relations

$$e_1 = r(\underline{x}) \cdot f_2$$

and $\quad e_2 = r(\underline{x}) \cdot f_1,$

where $r(\underline{x})$ is a function of set of variables, $\underline{x}$. The modulated gyrator preserves power; i.e., $P_1(t) = P_2(t)$.

3.3. Junction structure.

The junction structure of a bond graph is the set of all 0, 1, GY, and TF elements and their bonds and ports. The junction structure is an n-port that preserves power (i.e., the net power in is zero). The junction structure may be modulated (if it contains any MGY's or MTF's) or unmodulated.

For example, the junction structure of the graph in fig. 2(b) is a 4-port element with ports 1, 2, 5, and 6 and bonds 3 and 4. It contains the elements 0, TF, and 1.

4. PHYSICAL INTERPRETATIONS

The physical interpretations given in this section are very succinctly stated. References [1], [2], and [3] contain extensive descriptions of physical applications and the interested reader is encouraged to consult them.

4.1. Mechanical translation

To represent mechanical translational phenomena we may make the following variable associations.

(1) effort, e, is interpreted as force;

(2) flow, f, is interpreted as velocity;

(3) momentum, p, is interpreted as impulse-momentum;

(4) displacement, q, is interpreted as mechanical displacement.

Then the basic bond graph elements have the following interpretations:

(1) source of effort, SE, is a force source;

(2) source of flow, SF, is a velocity source (or may be thought of as a geometric constraint);

(3) resistance, R, represents friction and other mechanical loss mechanisms;

(4) capacitance, C, represents potential or elastic energy storage effects (or spring-like behavior);

(5) inertance, I, represents kinetic energy storage (or mass effects);

(6) transformer, TF, represents linear lever or linkage action (motion restricted to small angles);

(7) gyrator, GY, represents gyrational coupling or interaction between two ports;

(8) 0-junction represents a common force coupling among the several incident ports (or among the ports of the system bonded to the 0-junction); and

(9) 1-junction represents a common velocity constraint among the several incident ports (or among the ports of the system bonded to the 1-junction).

The extension of the interpretation to rotational mechanics is a natural one. It is based on the following associations:

(1) effort, e, is associated with torque; and

(2) flow, f, is associated with angular velocity.

Because the development is so similar to the one for translational mechanics it will not be repeated here.

4.2. Electrical networks

In electrical networks the key step is to interpret a port as a terminal-pair. Then variable associations may be made as follows:

(1) effort, e, is interpreted as voltage;

(2) flow, f, is interpreted as current;

(3) momentum, p, is interpreted as flux linkage;

(4) displacement, q, is interpreted as charge.

The basic bond graph elements have the following interpretations:

(1) source of effort, SE, is a voltage source;

(2) source of flow, SF, is a current source;

(3) resistance, R, represents electrical resistance;

(4) capacitance, C, represents capacitance effect (stored electric energy);

(5) inertance, I, represents inductance (stored magnetic energy);

(6) transformer, TF, represents ideal transformer coupling;

(7) gyrator, GY, represents gyrational coupling;

(8) 0-junction represents a parallel connection of ports (common voltage across the terminal pairs); and

(9) 1-junction represents a series connection of ports (common current through the terminal pairs).

4.3. Hydraulic circuits

For fluid systems in which the significant fluid power is given as the product of pressure times volume flow, the following variable associations are useful:

(1) effort, e, is interpreted as pressure;

(2) flow, f, is interpreted as volume flow;

(3) momentum, p, is interpreted as pressure-momentum;

(4) displacement, q, is interpreted as volume.

The basic bond graph elements have the following interpretations:

(1) source of effort, SE, is a pressure source;

(2) source of flow, SF, is a volume flow source;

(3) resistance, R, represents loss effects (e. g., due to leakage, valves, orifices, etc.);

(4) capacitance, C, represents accumulation or tank-like effects (head storage);

(5) inertance, I, represents slug-flow inertia effects;

(6) 0-junction represents a set of ports having a common pressure (e. g., a pipe tee);

(7) 1-junction represents a set of ports having a common volume flow (i. e., series).

4.4. Other interpretations

This brief listing of physical interpretations of bond graph elements is restricted to the simplest, most direct, applications. Such applications came first by virtue of historical development, and they are a natural point of departure for most classically-trained scientists and engineers. As references [1] - [4] and the special issue collection in the Journal of Dynamic Systems, Measurement, and Control, Trans. ASME, September 1972, indicate, bond graph elements can be used to describe an amazingly rich variety of complex dynamic systems. The limits of applicability are not bound by energy and power in the sense of physics; they include any areas in which there exist useful analogous quantities to energy.

5. CONCLUDING REMARKS

In this brief definition of the bond graph language two important concepts have been omitted. The first is the concept of bond activation, in which one of the two power variables is suppressed, producing a pure signal coupling in place of the bond. This is a very useful modeling device in active systems. Further discussion of activation will be found in reference (3), section 2.4, as well as in references [1] and [2].

Another concept omitted from discussion in this definitional paper is that of operational causality. It is by means of causality operations applied to bond graphs that the algebraic and differential relations implied by the graph and its elements may be organized and reduced to state-space form in a systematic manner. Extensive discussion of causality will be found in reference [3], section 3.4 and chapter 5. Systematic formulation of relations is presented in reference [6].

6. REFERENCES

[1] Paynter, H. M., "Analysis and Design of Engineering Systems," M.I.T. Press, 1961.

[2] Karnopp, D. C., and Rosenberg, R. C., "Analysis and Simulation of Multiport Systems," M.I.T. Press, 1968.

[3] Karnopp, D. C., and Rosenberg, R. C., "System Dynamics: A Unified Approach," Division of Engineering Research, College of Engineering, Michigan State University, East Lansing, Michigan, 1971.

[4] Takahaski, Y., Rabins, M., and Auslander, D., "Control," Addison-Wesley, Reading, Ma., 1970 (see Chapter 6 in particular).

[5] Rosenberg, R. C., "ENPORT User's Guide," Division of Engineering Research, College of Engineering, Michigan State University, East Lansing, Michigan, 1972.

[6] Rosenberg, R. C., "State-Space Formulation for Bond Graph Models of Multiport Systems," Trans. ASME, J. Dynamic Systems, March 1971, 35-40.

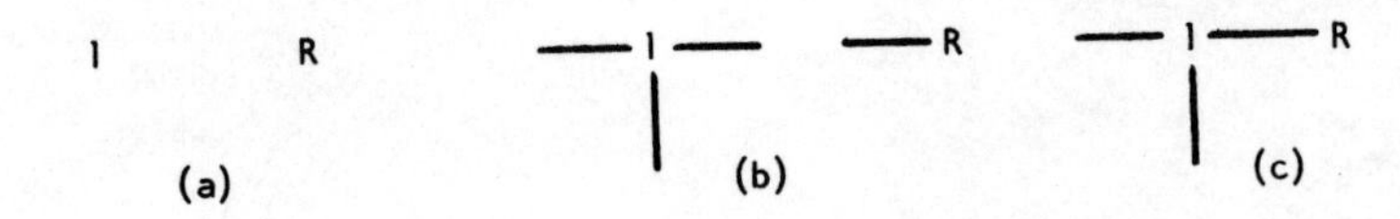

Figure 1. Multiport elements, ports, and bonds.
(a) two multiport elements
(b) the elements and their ports
(c) formation of a bond

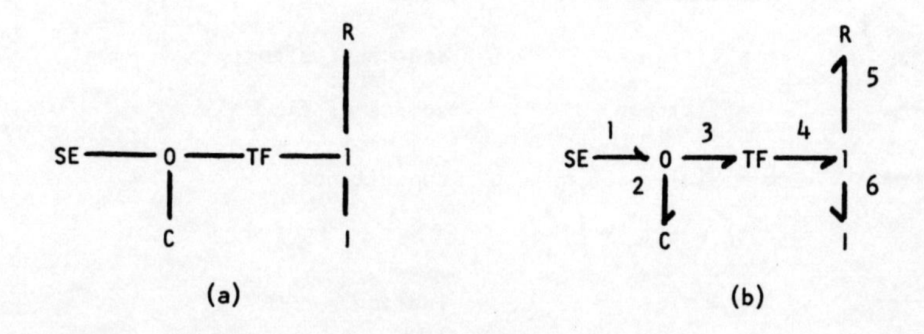

Figure 2. An example of a bond graph.
(a) a bond graph
(b) the bond graph with powers directed and bonds labeled

SYMBOL	DEFINITION	NAME
SE $\xrightarrow{e}$	$e = e(t)$	source of effort
SF $\xrightarrow{f}$	$f = f(t)$	source of flow
C $\leftharpoondown \frac{e}{f}$	$e = \Phi(q)$ $q(t) = q(t_o) + \int f \cdot dt$	capacitance
I $\leftharpoondown \frac{e}{f}$	$f = \Phi(p)$ $p(t) = p(t_o) + \int e \cdot dt$	inertance
R $\leftharpoondown \frac{e}{f}$	$\Phi(e,f) = 0$	resistance
$\xrightarrow{1}$ TF $\xrightarrow{2}$ 1:m	$e_1 = m \cdot e_2$ $m \cdot f_1 = f_2$	transformer
$\xrightarrow{1}$ GY $\xrightarrow{2}$ r	$e_1 = r \cdot f_2$ $e_2 = r \cdot f_1$	gyrator
$\xrightarrow{1}$ 0 $\xrightarrow{3}$, 2 $\uparrow$	$e_1 = e_2 = e_3$ $f_1 + f_2 - f_3 = 0$	common effort junction
$\xrightarrow{1}$ 1 $\xrightarrow{3}$, 2 $\uparrow$	$f_1 = f_2 = f_3$ $e_1 + e_2 - e_3 = 0$	common flow junction

Figure 3. Definitions of the basic multiport elements.

APPENDIX B. USER INFORMATION UPDATE SHEET

Return this sheet to:

Professor R.C. Rosenberg
Department of Mechanical Engineering
Michigan State University
East Lansing, Michigan 48824

if you wish to be kept informed of updates, revisions and interesting applications.

Name: ______________________________

Address: ______________________________

____

____

Date: ______________________________

I obtained my User's Guide from: ______________________________

____

I am ______(a) an ordinary user

______(b) in charge of a class/group using ENPORT-4

______(c) interested in/responsible for system maintenance

______(d) something else ______________________________

____

July, 1974

APPENDIX C. USER BUG AND USAGE SHEET

Dear Professor Rosenberg

1. I have found a bug in your program. I am not surprised, seeing that the program is so powerful and complex. Enclosed are my run(s) and comments. Briefly, the error is:

2. I have found a terrific application of your program. I'll bet you didn't know this could be done. Enclosed are details. Briefly, the idea is:

3. I wish to inform you of the following fascinating observation/fact/ conjecture/rumor/insight/ ____________:

July, 1974

APPENDIX D

Current Size Limitations in ENPORT-4

1)	Bonds	total maximum	=	65
		external	=	50
		internal	=	15
2)	Multiport elements maximum		=	65
3)	Number of ports C, I, R maximum		=	5
		0, 1 maximum	=	5
4)	Independent energy variables		=	20
5)	Dependent energy variables		=	10
6)	Number of sources maximum		=	10
7)	Number of resistances maximum		=	20

APPENDIX E

Error Codes for ENPORT ver 4.0

Number	Definition
100	Error in program structure, Level-1 command at this point must be GRAPH.
101	More than five bonds attached to a single node.
102	Maximum number of nodes exceeded; see appendix for appropriate program limits.
103	Illegal node name, must be C, I, R, SE, SF, TF, GY, 0 or 1.
104	Maximum number of bonds exceeded; see appendix for appropriate program limits.
105	Bond "i" connected to more than two nodes. "i" = ABORT
106	Bond numbering not sequential.
107	Bond "i" attached at only one end. "i" = ABORT
108	Power conflict in assignment at node "i". "i" = ABORT
109	Illegal control word, must be ASSIGN, CAUSAL, ACTIVATE, or GREND.
110	Bond "i" cannot be assigned as either internal or external. "i" = ABORT
111	S4 matrix (internal vs internal) is singular.
112	FAR or P4 matrix is singular, illegal R-field parameters.
113	S33 matrix is singular.
114	F22 matrix is singular.
115	WS11 matrix is singular.
116	WS11 matrix is singular.
117	Invalid output request on non-energy bond.
118	Illegal level-1 command; see users manual Chapter 2.

Number	Definition
119	Illegal source variable.
120	Illegal setup directive.
121	Unable to find suitable dt for matrix expansion.
122	MSS matrix is singular.
123	Bond number given is greater than allowed maximum.
124	Error in number field of output request.
125	Invalid output request.
126	Maximum number of internal bonds exceeded.
127	Maximum number of external bonds exceeded.
128	Maximum number of energy bonds exceeded.
129	Maximum number of dependent energy bonds exceeded.
130	Maximum number of dissipation bonds exceeded.
131	Maximum number of source bonds exceeded.
132	Illegal source definition (type).
133	Activation with level-3 causality requires user supplied causality assignments.
134	Causal assignment error on 0- or 1-junction.
135	Node name given not attached to bond number given.
136	Bonds must be assigned either an E or F.
137	Causal assignment error on a TF or GY.
138	Illegal source type for dependent field problem.
139	Illegal output request.
140	Illegal output variable request.
141	Illegal list terminator, must be a ".".
142	Illegal node name in parameter string.
143	Unable to match parameter identifier.
144	Error in parameter value field.

Number	Definition
145	PREND card missing or duplicate parameters.
146	Illegal Power request on bond"i". "I" = ABORT
147	Missing period on graph string or illegal node name.
148	W1 matrix is singular.
149	W6 matrix is singular.

APPENDIX F. Causality Assignment Procedures

Level 0 : User-supplied causal information (see CAUSAL, section 2.1.2), if any, is processed, with causal implications extended as far as possible through the graph structure.

Level 1 : The node list is searched from top to bottom for SE or SF elements, and appropriate causality is assigned. As each SE or SF is assigned, causality is extended.

Level 2 : The node list is searched from top to bottom for C or I elements, and integration causality is assigned. Causality is extended after each element assignment.

Level 3 : After the previous information has been processed, the node list is searched from top to bottom for R elements. R elements are assigned such that their input is a flow. Causality is extended after each element assignment.

Level 4 : The program stops if this level is reached.

Remarks :
1. Causality assignment may terminate at any level, once the graph is completely determined.
2. Only the first bond on an R-field (in order of the line code) may be set causally by the user. The other bonds will be given the same causality, if that is possible.
3. Graph causality can be controlled partially through input node order, as the node list is searched from top to bottom. This is particularly relevant for interacting C and I elements.

APPENDIX G

ENPORT-Ver 4.0 System Structure

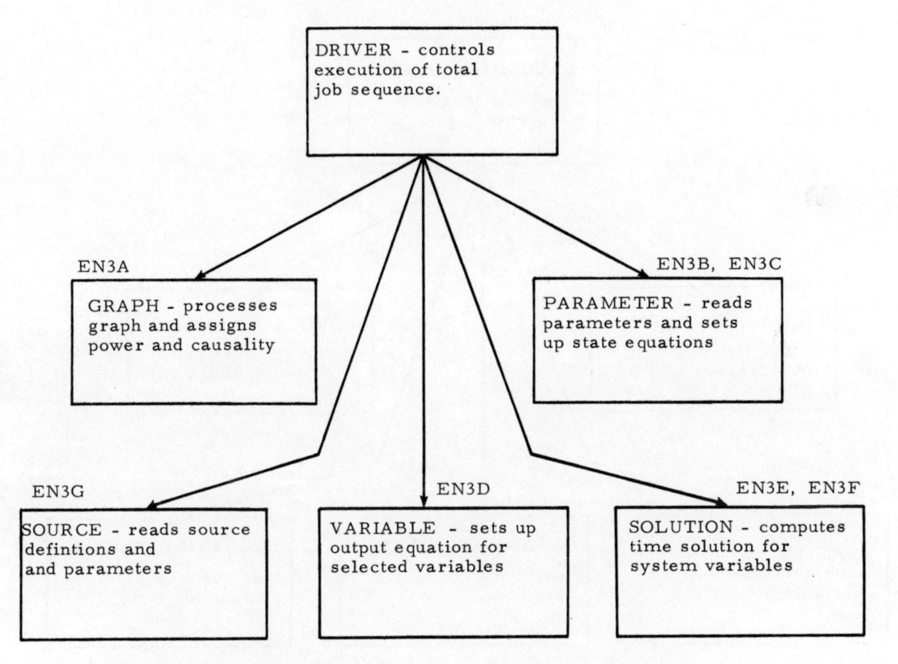

Notes:

1. Driver reads all of the input control deck and calls each of the subsections as needed.
2. Output printing and plotting are provided as part of the solution subsection.
3. Block references on succeeding pages are to Figure 4-1.

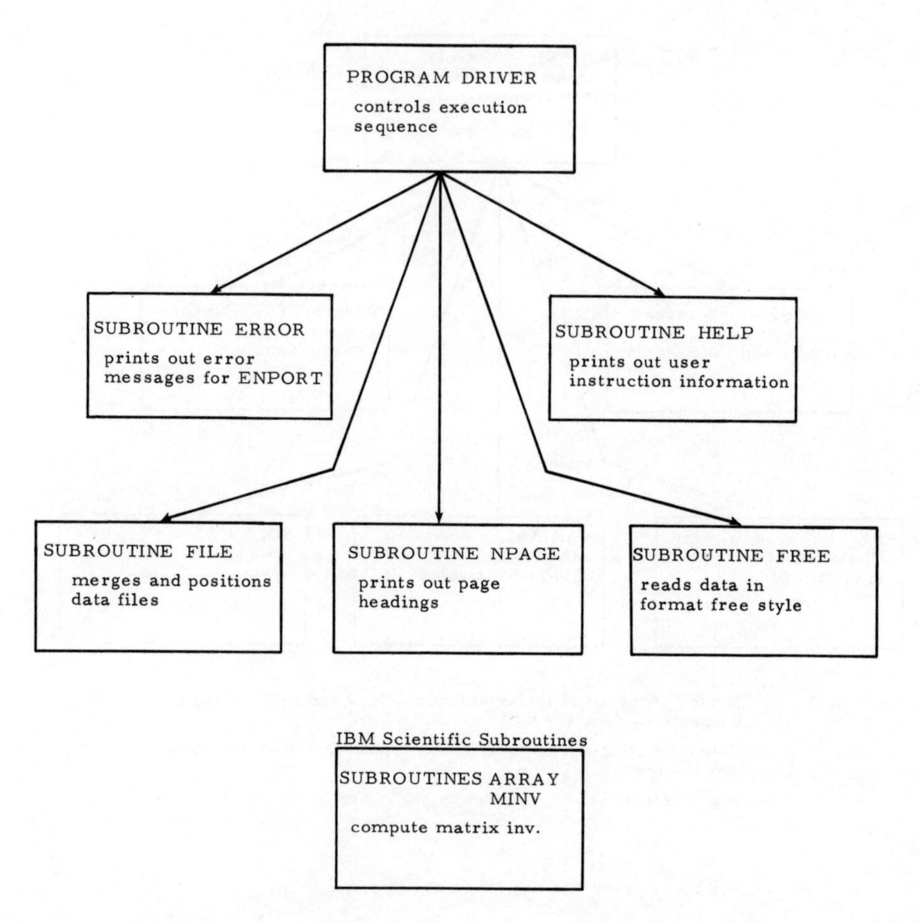

IBM Scientific Subroutines

SUBROUTINES ARRAY
MINV
compute matrix inv.

General purpose functions.

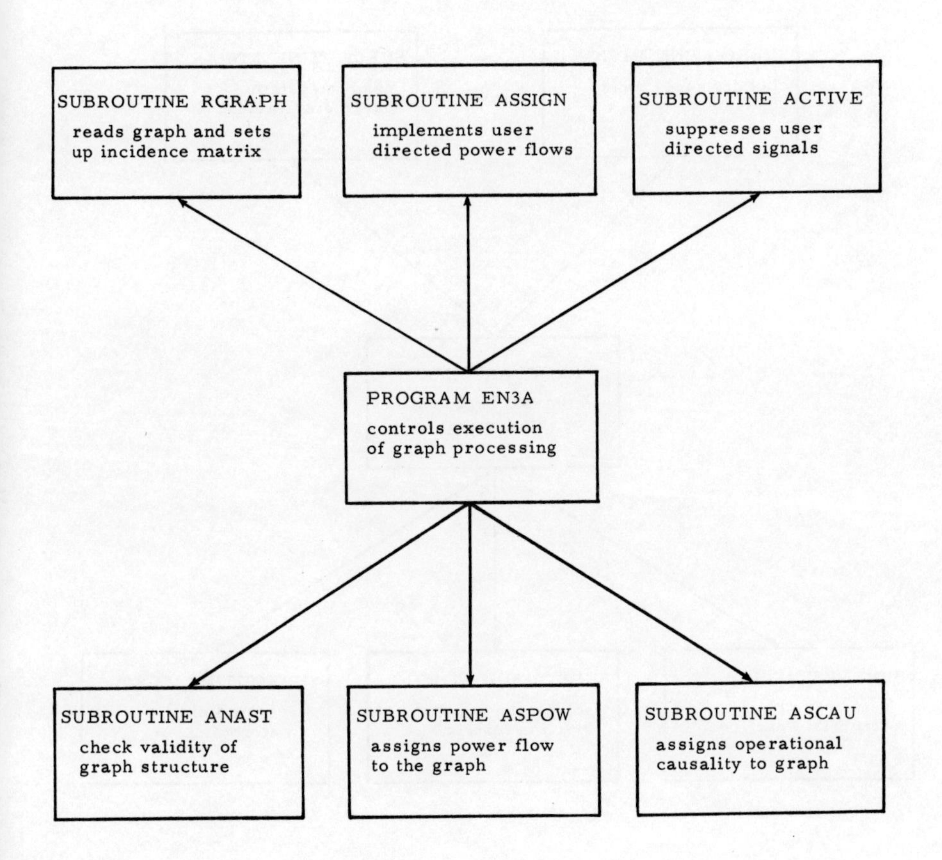

GRAPH Processing functions (blocks 1, 2 and 3).

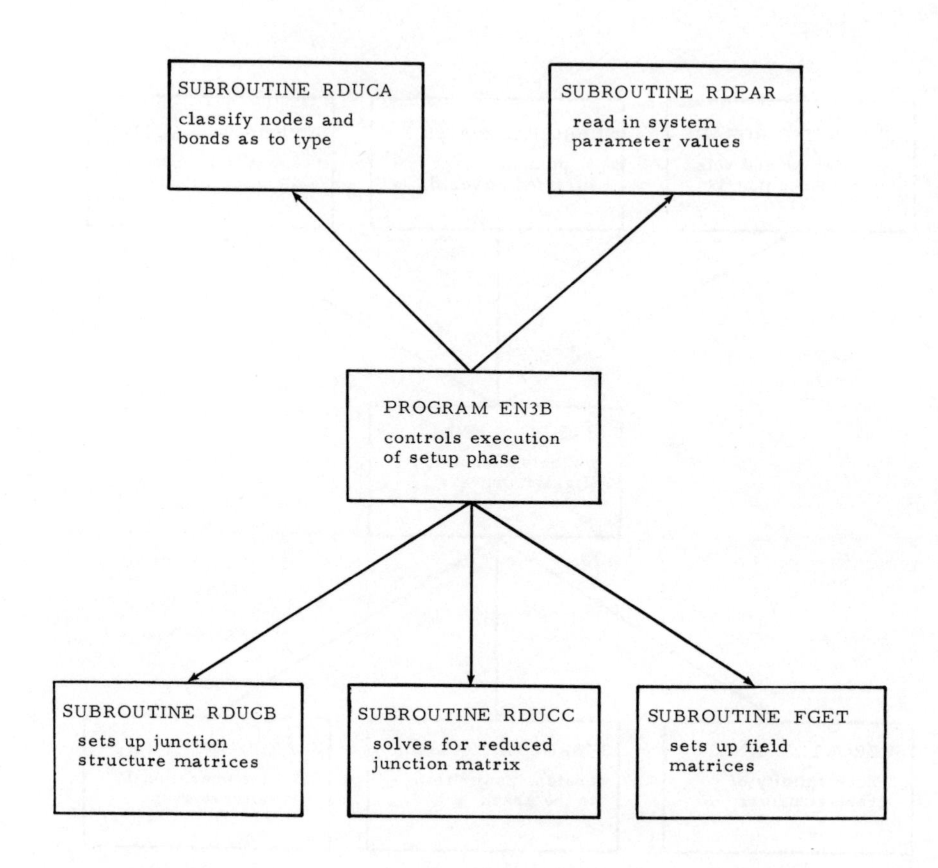

Classification and initial setup phase (blocks 4 and 5).

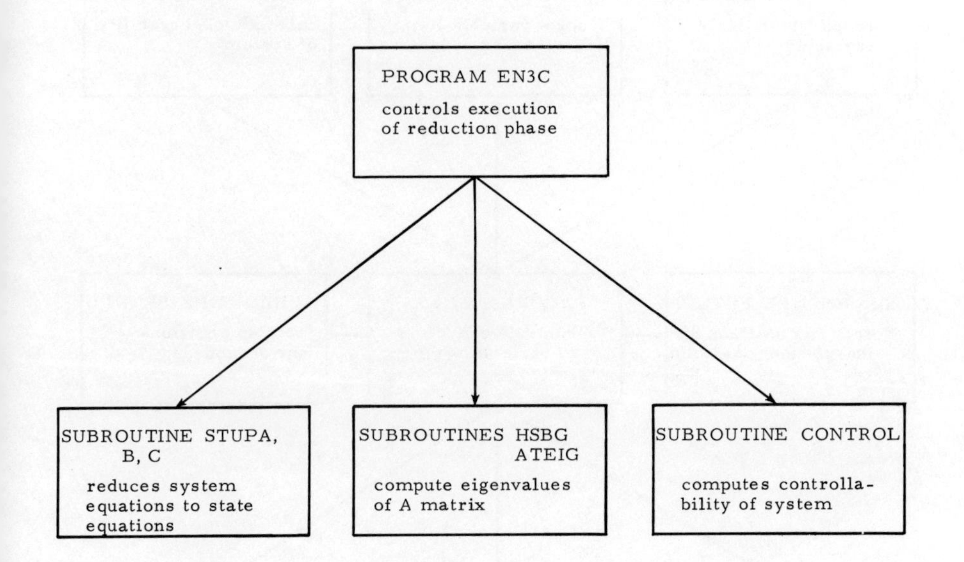

Reduction phase (blocks 6, 7, 8 and 9).

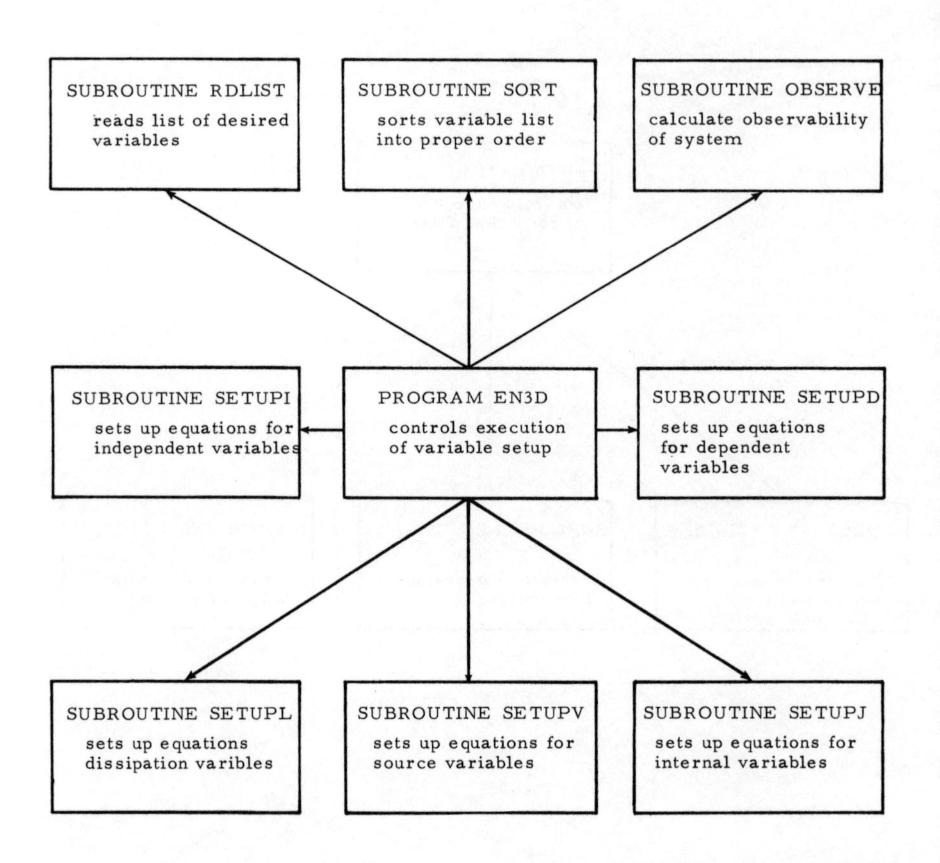

SETUP function (blocks 10 and 11).

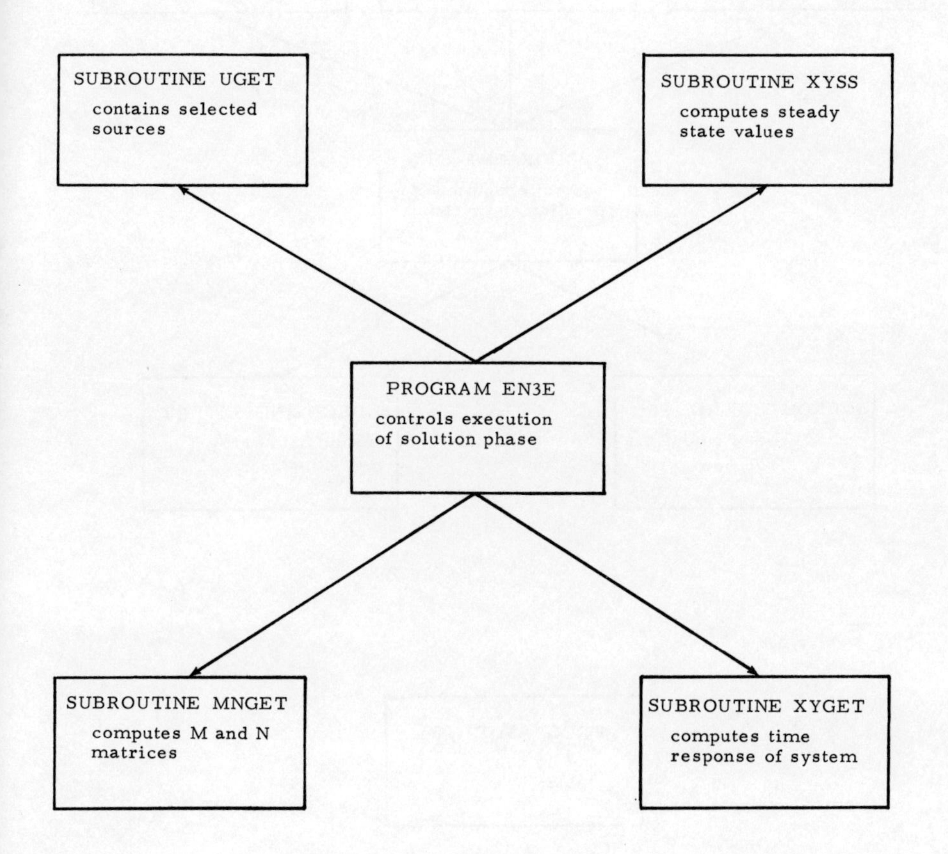

Solution phase (blocks 12 and 13).

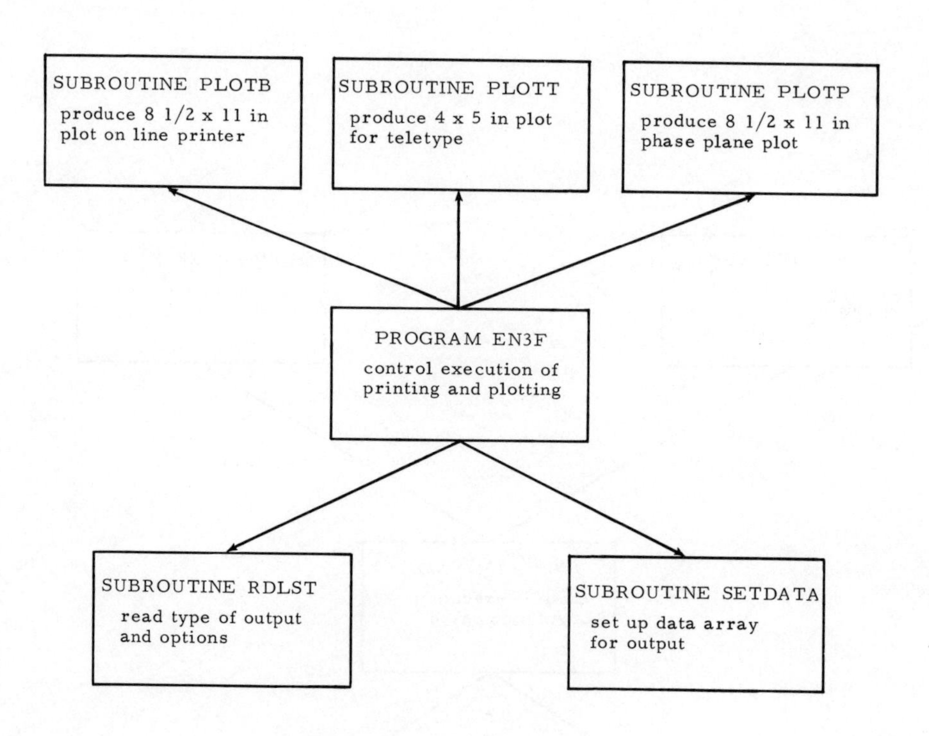

PROGRAM EN3G

set up source tables

Output functions (block 14).

ENPORT Ver 4.0 UTILITY FILES

Tape Number	Description
1.	Internal data file; Primary working storage area for graph being processed. The file contains graph definition matrices, parameters, source definition matrices, state and output equation matrices and variable time responses.
2.	System input file; BATCH - ENPORT job deck is read on this file and copied to tape 7. INTERACTIVE - Teletype keyboard input.
3.	System output file; Results of ENPORT run (batch and interactive) are written on this file for printing.
4.	Variables backup file; Contains copy of output equation matrices and variable list.
5.	Utility output file; BPLOT's and PPLOT's which are called for from the interactive mode are written on this file for disposal to the line printer.
6.	Utility punch file; System time response data written in punchable format in this file.
7.	Batch input file; The ENPORT job deck is copied to this file from tape 2 for program processing.
8.	Sources backup file; Contains source definitions and source parameters.
9.	Program library file; Program library facilities not yet implemented.